Student Solutions Manual

Numerical Mathematics and Computing

AF898806

SEVENTH EDITION

E. Ward Cheney
University of Texas at Austin

David R. Kincaid
University of Texas at Austin

BROOKS/COLE
CENGAGE Learning

Australia • Brazil • Japan • Korea • Mexico • Singapore • Spain • United Kingdom • United States

BROOKS/COLE
CENGAGE Learning

ISBN-13: 978-1-133-49180-4
ISBN-10: 1-133-49180-4

Brooks/Cole
20 Channel Center Street
Boston, MA 02210
USA

Cengage Learning is a leading provider of customized learning solutions with office locations around the globe, including Singapore, the United Kingdom, Australia, Mexico, Brazil, and Japan. Locate your local office at: **www.cengage.com/global**

Cengage Learning products are represented in Canada by Nelson Education, Ltd.

To learn more about Brooks/Cole, visit **www.cengage.com/brookscole**

Purchase any of our products at your local college store or at our preferred online store **www.cengagebrain.com**

Printed in the USA
1 2 3 4 5 30 29 28 27 26

10 Monte Carlo Methods and Simulation **143**
10.1 Random Numbers 143
10.1 Computer Exercises 143
10.2 Estimation of Areas and Volumes by Monte Carlo Techniques 143
10.2 Computer Exercises 144
10.3 Simulation: Computer Exercises 144

11 Boundary-Value Problems **145**
11.1 Shooting Method 145
11.1 Computer Exercises 147
11.2 A Discretization Method 148
11.2 Computer Exercises 149

12 Partial Differential Equations **150**
12.1 Parabolic Problems 150
12.2 Hyperbolic Problems 152
12.2 Computer Exercises 153
12.3 Elliptic Problems 153
12.3 Computer Exercises 155

13 Minimization of Functions **156**
13.1 One Variable Case 156
13.2 Multivariate Case 158

14 Linear Programming Problems **161**
14.1 Standard Forms and Duality 161
14.1 Computer Exercises 165
14.2 Simplex Method 166
14.2 Computer Exercises 166
14.3 Inconsistent Linear Systems 167
14.3 Computer Exercises 168

A Upper and Lower Sums **169**

B Representation of Numbers in Different Bases **177**

C MATLAB: Brief Introduction and Tips **179**

D Maple: Brief Introduction and Tips **186**

E Mathematica: Brief Introduction and Tips **192**

F Mathematical Software Tasks **197**

G Reviews for Exams **198**

H Lists of Sample Cod es **201**

Chapter 1

Mathematical Preliminaries and Floating-Point Representation

1.1 Introduction

1.1.2. $(x - 0.6032)/x = 1/1000;\ 1000x - 0.6032 = 603.2;\ x = 6032/9990$

$(0.6032 - x)/x = 1/1000;\ 603.2 - 1000x = x;\ x = 6032/10010$

1.1.3. Relative Error $= \dfrac{|5.000 - 4.9997|}{4.9997} \approx 6 \times 10^{-5}$

1.1.4. $pi \leftarrow 2.0 \arcsin(1.0)$

$pi \leftarrow 2.0 \arccos(0.0)$

$pi \leftarrow \arccos(-1.0)$

$pi \leftarrow -2.0 \arcsin(-1.0)$

1.1.5. **a.**

```
sum ← 0.0
for i = 1 to n
   for j = 1 to n
      sum ← sum + a_ij
   end for
end for
```

d.

```
sum ← 0.0
for i = 1 to n
   sum ← sum + a_ii
end for
for j = 2 to n
   for i = j to n
      sum ← sum + a_{i,i-j+1} + a_{i-j+1,i}
   end for
end for
```

1.1.6. By the pseudocode in the textbook, it is clear that one addition and one multiplication occurs for each step in the loop. Since there are n steps, we have n additions/subtractions and n multiplications.

1.1.8. **a.** $p(x) = x^{2^5}$

for $i = 1$ **to** 5
  $x \leftarrow x \cdot x$
end for
$p \leftarrow x$

c. $p(x) = (x+2)^3[6 + 9(x+2)^4 + 3(x+2)^{12} - (x+2)^{28}]$
$= (x+2)^3[6 + (x+2)^4[9 + (x+2)^8[3 - (x+2)^{16}]]]$
$z \leftarrow x + 2$; $p \leftarrow z^3(6 + z^4(9 + z^8(3 - z^{16})))$

1.1.10. $\sum_{i=1}^{n} b_i^{-1} \prod_{j=1}^{i} a_j$
$= b_1^{-1}a_1 + b_2^{-1}a_1a_2 + b_3^{-1}a_1a_2a_3 + \cdots + b_{n-1}^{-1}(a_1 \cdots a_{n-1}) + b_n^{-1}(a_1 \cdots a_n)$

$z \leftarrow a_n/b_n$
for $i = 1$ **to** $n - 1$
  $z \leftarrow a_{n-i}(z + 1/b_{n-i})$
end for

Successive contents of cell z:

$$
\begin{aligned}
z &= b_n^{-1}a_n \\
z &= a_{n-1}(b_n^{-1}a_n + b_{n-1}^{-1}) = b_n^{-1}a_na_{n-1} + b_{n-1}^{-1}a_{n-1} \\
z &= a_{n-2}(b_n^{-1}a_na_{n-1} + b_{n-1}^{-1}a_{n-1} + b_{n-2}^{-1}) \\
&= b_n^{-1}(a_na_{n-1}a_{n-2}) + b_{n-1}^{-1}(a_{n-1}a_{n-2}) + b_{n-2}^{-1}a_{n-2}
\end{aligned}
$$

1.1.11. **b.**

$z \leftarrow 1$
$v \leftarrow 1$
for $i = 1$ **to** $n - 1$
  $v \leftarrow vx$
  $z \leftarrow vz + 1$
end for
$z \leftarrow vxz$

1.1.12. **b.** $v = \sum_{i=0}^{n} a_i x^i$;
$(\cdots((a_nx + a_{n-1})x + a_{n-2})x + \cdots + a_1)x + a_0$

e. $v = a_nx^n + x\sum_{i=1}^{n} a_{n-i}x^{n-i}$;
$(\cdots((a_n + a_{n-1})x + a_{n-2})x + \cdots + a_0)x$
$= a_nx^n + a_{n-1}x^n + a_{n-2}x^{n-1} + \cdots + a_0x$

1.1.13. $(\cdots(((b_n + 1)b_{n-1} + 1)b_{n-2} + 1)\cdots)b_2 + 1 = b_nb_{n-1}\cdots b_2 + b_{n-1}b_{n-2}\cdots b_2 + \cdots + b_2 + 1$

$$z = 1 + \sum_{1=2}^{n} \prod_{j=2}^{i} b_j$$

1.1.14.

j-value	1	2	3	$\cdots$	n
# mult.	1	2	3	$\cdots$	n

Total # mult. $= 1 + 2 + 3 + \cdots + n = n(n+1)/2$

1.1.15. **b.** The values in the A-array are assigned down the columns.

```
for j = 1 to n
   for i = 1 to n
      a_ij ← 1.0/ real(i + j − 1)
   end for
end for
```

1.1 Computer Exercises

1.1.C4. $\exp(1.0) \approx 2.71828\ 18284\ 6$.

1.1.C9. All a_i should be equal to a_1. However, computation deviates from theory when $a_1 = 10^{-12}$, 10^{-8}, 10^{-4}, 10^{20}, for example.

1.1.C10. The value of x may underflow and be set to zero by the computer.

1.1.C12. Forty different spellings.

```
character lpart(4), rpart(5)
integer i, j
lpart(1) ← 'CHEB'
lpart(2) ← 'CEB'
lpart(3) ← 'TSCHEB'
lpart(4) ← 'TCHEB'
rpart(1) ← 'SHEV'
rpart(2) ← 'SCHEF'
rpart(3) ← 'CEV'
rpart(4) ← 'CHEFF'
rpart(5) ← 'SCHEFF'
for i = 1 to 4
   for j = 1 to 5
      output lpart(i), 'Y', rpart(j)
      output lpart(i), 'I', rpart(j)
   end for
end for
```

1.1.C20. **a.** The computation real(m/n) may result in truncation so that $x \neq y$.

1.2 Mathematical Preliminaries

1.2.4. **a.** First derivative unbounded:

$$f(x) = x^{1/2}, \qquad f(0) = 0$$
$$f'(x) = \frac{1}{2}x^{-1/2}, \quad f'(0) = \infty$$

b. First derivative not continuous:

$$f(x) = |x|$$
$$f'(x) = \begin{cases} 1, & x > 0 \\ 0, & x = 0 \\ -1, & x < 0 \end{cases}$$

e. Function not defined at 0:

$$f(x) = \log x \quad f(0) = -\infty$$

1.2.5. $f(x) = f(a) + (x-a)f'(a) + (x-a)^2 f''(a)/2! + \cdots$
replace a with $x-h$
$f(x) = f(x-h) + hf'(x-h) + h^2 f''(x-h)/2! + \cdots$
replace x with $x+h$
$f(x+h) = f(x) + hf'(x) + h^2 f''(x)/2! + \cdots$

$$\begin{aligned} f(x) &= \cosh x, & f(0) &= 1 \\ f'(x) &= \sinh x, & f'(0) &= 0 \\ f''(x) &= \cosh x, & f''(0) &= 1 \\ f'''(x) &= \sinh x, & f'''(0) &= 0 \end{aligned}$$

$f(x) = f(a) + (x-a)f'(a) + (x-a)^2 f''(a)/2! + \cdots$

$$\cosh x = 1 + \frac{x^2}{2!} + \frac{x^4}{4!} + \frac{x^6}{6!} + \cdots = \sum_{k=0}^{\infty} \frac{(x)^{2k}}{(2k)!}$$

$$\begin{aligned} \cosh 0.7 \approx\ & 1.0 \\ & 0.245 \\ & 0.010004167 \qquad \cosh 0.7 \approx 1.25517 \\ & \underline{0.000163401} \\ & 1.255167568 \end{aligned}$$

1.2.6. **a.**

$$\begin{aligned} f(x) &= e^{\cos x}, & f(0) &= e \\ f'(x) &= e^{\cos x}(-\sin x), & f'(0) &= 0 \\ f''(x) &= e^{\cos x}(\sin^2 x) + e^{\cos x}(-\cos x), & f''(0) &= -e \end{aligned}$$

$$\begin{aligned} f(x) &= f(0) + xf'(0) + \frac{x^2}{2!}f''(0) + \cdots \\ &= e + 0 - \frac{x^2}{2}e + \cdots \end{aligned}$$

Thus, we obtain $e^{\cos x} = e\Big(1 - \dfrac{x^2}{2} + \cdots\Big)$

b.

$$\begin{aligned} f(x) &= \sin(\cos x), & f(0) &= \sin 1 \\ f'(x) &= \cos(\cos x)(-\sin x), & f'(0) &= 0 \\ f''(x) &= -\sin(\cos x)(\sin^2 x) - \cos(\cos x)\cos x, & f''(0) &= -\cos 1 \end{aligned}$$

$$\sin(\cos x) = (\sin 1) - (\cos 1)\left(\frac{x^2}{2}\right) + \cdots$$

1.2.7.

$$\begin{aligned} f(x) &= (x-1)^{1/2}, & f(m) &= (m-1)^{1/2} \\ f'(x) &= \frac{1}{2}(x-1)^{-1/2}, & f'(m) &= \frac{1}{2}(m-1)^{-1/2} \\ f''(x) &= -\frac{1}{4}(x-1)^{-3/2}, & f''(m) &= -\frac{1}{4}(m-1)^{-3/2} \\ &\vdots & &\vdots \end{aligned}$$

Clearly, $(m-1) > 0$ for $f^{(k)}(m)$ to exist. Hence, we obtain $m = 2$.

1.2.8.

$$\begin{aligned} f(x) &= e^x, & f(0) &= 1 \\ f'(x) &= e^x, & f'(0) &= 1 \\ f''(x) &= e^x, & f''(0) &= 1 \\ &\vdots & &\vdots \end{aligned}$$

$$f(x) = f(0) + xf'(0) + \frac{x^2}{2} f''(0) + \cdots$$

Therefore, we obtain $e^x = 1 + x + \frac{x^2}{2!} + \frac{x^3}{3!} + \frac{x^4}{4!} + \cdots$ Now

$$e = 1 + 1 + \frac{1}{2!} + \frac{1}{3!} + \frac{1}{4!} + \cdots$$

By Taylor's Theorem, if we stop at $1/(n-1)!$ the error is $e^\xi/n!$ where $0 < \xi < 1$.

So we require: $\frac{e}{n!} < \frac{1}{2}10^{-15}$

$$\Rightarrow \log e - \log n! < \log\frac{1}{2} - 15$$

Therefore, we have $15 + \log(2e) < \log(n!)$

$$\left.\begin{aligned} n = 17: &\quad 15.74 \not< 14.55 \\ n = 18: &\quad 15.74 < 15.81 \end{aligned}\right\} \Rightarrow 18 \text{ terms}$$

1.2.9. We have $e^{-x} = 1 - x + x^2/2! - x^3/3! + x^4/4! - \cdots$ in which the signs alternate.

$$e^{-x} = 1/e^x = 1\Big/\left(1 + x + \frac{x^2}{2!} + \frac{x^3}{3!} + \frac{x^4}{4!} + \cdots\right)$$

Yes, this method is practical. By using this formula, we avoid the series for e^{-x} and use the one for e^x.

1.2.11.

$$\begin{aligned}
\ln(1+x) &= x - \frac{x^2}{2} + \frac{x^3}{3} - \frac{x^4}{4} + \cdots \\
\ln(1-x) &= -x - \frac{x^2}{2} - \frac{x^3}{3} - \frac{x^4}{4} - \cdots \\
\ln\left(\frac{1+x}{1-x}\right) &= \ln(1+x) - \ln(1-x) \\
&= 2\left[x + \frac{x^3}{3} + \frac{x^5}{5} + \cdots\right] = 2\sum_{k=1}^{\infty} \frac{x^{2k-1}}{(2k-1)}
\end{aligned}$$

1.2.12.

$$\left(\frac{1+x}{1-x}\right) = 2 \Rightarrow (1+x) = 2(1-x) \Rightarrow x = \frac{1}{3}$$

Therefore, we have

$$\ln\ 2 = \ln\left(\frac{1+\frac{1}{3}}{1-\frac{1}{3}}\right) = 2\left[3^{-1} + \frac{3^{-3}}{3} + \frac{3^{-5}}{5} + \frac{3^{-7}}{7} + \cdots\right] \approx 0.69313 \text{ (4 terms)}$$

If we take 10 terms and denote the error by E, then

$$\begin{aligned}
\ln\ 2 &= 2\left[3^{-1} + \frac{3^{-3}}{3} + \cdots + \frac{3^{-19}}{19}\right] + E \\
E &= 2\left[\frac{3^{-21}}{21} + \frac{3^{-23}}{23} + \cdots\right] \\
&< \frac{2}{20}3^{-21}\left[1 + \frac{1}{9} + \left(\frac{1}{9}\right)^2 + \cdots\right] = \frac{2}{20}3^{-21}\frac{1}{1-1/9} \\
&\approx 1.08 \times 10^{-11}
\end{aligned}$$

Yes, this method is practical.

1.2.15.

$$\begin{aligned}
\sin x &= x - \frac{x^3}{3!} + \frac{x^5}{5!} - \frac{x^7}{7!} + \cdots \\
\cos x &= 1 - \frac{x^2}{2!} + \frac{x^4}{4!} - \frac{x^6}{6!} + \cdots
\end{aligned}$$

a. Now we have

$$\sin x + \cos x = 1 + x - \frac{x^2}{2} - \frac{x^3}{6} + \cdots$$

Then

$$\begin{aligned}
\sin(.001) + \cos(.001) &\approx 1 + .001 - .0000005 - .00000000017 \\
\sin(.001) + \cos(.001) &\approx 1.00099\,94998\,3
\end{aligned}$$

b.

$$
\begin{aligned}
\sin x \cos x = x &- \frac{x^3}{6} + \frac{x^5}{120} - \frac{x^7}{5040} \\
&- \frac{x^3}{2} + \frac{x^5}{12} - \frac{x^7}{240} \\
&\qquad\quad + \frac{x^5}{24} - \frac{x^7}{144} \\
&\qquad\qquad\qquad - \frac{x^7}{720} + \cdots \\
\sin x \cos x = x &- \frac{2}{3}x^3 + \frac{2}{15}x^5 - \frac{4}{315}x^7 + \cdots
\end{aligned}
$$

$\sin(.0006)\cos(.0006) \approx 0.0006 - 1.44 \times 10^{-10} + 1.04 \times 10^{-12}$

$\sin(.0006)\cos(.0006) \approx 0.00059\,99998\,57$

1.2.16.

$$f(x) = f(0) + xf'(0) + \frac{x^2}{2}f''(0) + \frac{x^2}{3!}f'''(0) + \cdots$$

$$
\begin{aligned}
\text{Let} \quad f(x) &= \ln(e+x), & f(0) &= 1 \\
f'(x) &= (e+x)^{-1}, & f'(0) &= e^{-1} \\
f''(x) &= -(e+x)^{-2}, & f''(0) &= -e^{-2} \\
f'''(x) &= 2(e+x)^{-3}, & f'''(0) &= 2e^{-3} \\
f^{(4)}(x) &= -3!(e+x)^{-4}, & f^{(4)}(0) &= -3!e^{-4}
\end{aligned}
$$

Thus, we have $\ln(e+x) = 1 + \frac{x}{e} - \frac{x^2}{2e^2} + \frac{x^3}{3e^3} - \frac{x^4}{4e^4} + \cdots$

$$\ln(e+x) = 1 + \sum_{n=1}^{\infty} (-1)^{n-1}\frac{1}{n}\left(\frac{x}{e}\right)^n$$

$$
\begin{aligned}
\text{Check: } \ln(e+x) - 1 &= \sum_{n=1}^{\infty} (-1)^{n-1}\frac{1}{n}\left(\frac{x}{e}\right)^n \\
\ln(e+x) - \ln\ e &= \ln\left(1 + \frac{x}{e}\right)
\end{aligned}
$$

1.2.17.

$$(-1)^n\frac{x^{2n}}{(2n)!} \quad (n+1) - \text{term}, \quad |x| < \frac{1}{2}$$

$$|\text{Error}| \le \left|\frac{x^{2n}}{(2n)!}\right| < \frac{2^{-2n}}{(2n)!} < \frac{1}{2}10^{-12}$$

$$\frac{2^{-2n+1}}{(2n)!} < 10^{-12}$$

$$(-2n+1)\log 2 - \log(2n)! < -12$$

$$\left.\begin{aligned} n = 6: \quad \text{LHS} &= -11.99 \\ n = 7: \quad \text{LHS} &= -14.85 \end{aligned}\right\} \Rightarrow 7 \text{ terms}$$

1.2.18.

$$f(x) = \sum_{k=1}^{\infty}(-1)^k\left(\frac{x^k}{k^4}\right) = -x + \frac{x^2}{2^4} - \frac{x^3}{3^4} + \cdots$$

$$f(1) = -1 + \frac{1}{2^4} - \frac{1}{3^4} + \cdots + \frac{(-1)^n}{n^4} + \cdots$$

$$\text{want } \frac{1}{(n+1)^4} < 10^{-8}$$

$$\Rightarrow (n+1)^4 > 10^8$$

$$\Rightarrow n > 10^2 - 1 \quad \Rightarrow \quad 100 \text{ terms}$$

1.2.20.

$$\begin{aligned} f(x) &= x^{1/2}, & f(1) &= 1 \\ f'(x) &= \frac{1}{2}x^{-1/2}, & f'(1) &= \frac{1}{2} \\ f''(x) &= -\frac{1}{2^2}x^{-3/2}, & f''(1) &= -\frac{1}{4} \\ f'''(x) &= \frac{3}{2^3}x^{-5/2}, & f'''(1) &= \frac{3}{8} \\ f^{(4)}(x) &= \frac{3\cdot 5}{2^4}x^{-7/2}, & f^{(4)}(1) &= -\frac{15}{16} \end{aligned}$$

$$f(x+h) = f(x) + hf'(x) + \frac{h^2}{2}f''(x) + \frac{h^3}{3!}f'''(x) + \frac{h^4}{4!}f^{(4)}(x)$$

$$f(1-2h) = f(1) - 2hf'(1) + 2h^2f''(1) - \frac{4}{3}h^3f'''(1) + \frac{2h^4}{3}f^{(4)}(1)$$

$$= 1 - h - \frac{h^2}{2} - \frac{h^3}{2} - \frac{5}{8}h^4$$

$$\text{Fifth term is } -\frac{5}{8}h^4$$

1.2.23.

$$f = (4x - z)^{1/4}, \qquad f\left(\tfrac{17}{4}\right) = 2$$

$$f' = (4x-1)^{-3/4}, \qquad f'\left(\tfrac{17}{4}\right) = \frac{1}{8}$$

Since $f(x) = f(a) + (x-a)f'(a) + \cdots$, the second term is $\frac{1}{8}\left(x - \frac{17}{4}\right)$.

1.2.24.

$$\log n! = \log 1 + \log 2 + \log 3 + \cdots + \log n$$

```
s ← 0
for i = 2 to n
  s ← s + log(i)
  output i, s
end for
```

1.2.28.

$$\left|\cos x - \left(1 - \frac{x^2}{2}\right)\right| \le \frac{x^4}{4!} < \frac{1}{16 \times 24} = \frac{1}{384}$$

1.2.32. Maclaurin series:

$$\begin{aligned}
f(x) &= f(0) + xf'(0) + \frac{x^2}{2!}f''(0) + \frac{x^3}{3!}f'''(0) + \frac{x^4}{4!}f^{(4)}(0) + \cdots \\
&= 3 + x(7) + \frac{x^2}{2!}(-2.66) + \frac{x^3}{3!}(0) + \frac{x^4}{4!}(460.8) \\
&= 3 + 7x - 1.33x^2 + 19.2x^4 \\
f(x) &= 318.88 + (x-2)616.08 + \frac{(x-2)^2}{2!}918.94 + \frac{(x-2)^3}{3!}921.6 + \frac{(x-2)^4}{4!}460.8
\end{aligned}$$

1.2.35. $\dfrac{4}{(2n-1)}$ is nth term; $\dfrac{4}{2n+1} < \dfrac{1}{2}10^{-2}$; $2n+1 > 800$; $n > 399.5$; 400 terms.

1.2.38.

$$\begin{aligned}
f(x) &= \cos x, & f(60^\circ) &= \frac{1}{2} \\
f'(x) &= -\sin x, & f'(60^\circ) &= -\frac{\sqrt{3}}{2} \\
f''(x) &= -\cos x, & f''(60^\circ) &= -\frac{1}{2} \\
f'''(x) &= \sin x, & f'''(60^\circ) &= \frac{\sqrt{3}}{2} \\
&\vdots & &\vdots
\end{aligned}$$

$$f(x+h) = f(x) + hf'(x) + \frac{h^2}{2}f''(x) + \frac{h^3}{3!}f'''(x) + \cdots$$

$$\cos\left(\frac{\pi}{3} + h\right) = \frac{1}{2} + h\left(-\frac{\sqrt{3}}{2}\right) + \frac{h^2}{2}\left(-\frac{1}{2}\right) + \frac{h^3}{3!}\left(\frac{\sqrt{3}}{2}\right) + \cdots$$

$$\cos\left(\frac{\pi}{3} + h\right) = \frac{1}{2}\left(1 - \frac{h^2}{2!} + \frac{h^4}{4!} - \frac{h^6}{6!} + \cdots\right) + \frac{\sqrt{3}}{2}\left(-h + \frac{h^3}{3!} - \frac{h^5}{5!} + \cdots\right)$$

$$h = 10^{-3} \cdot \frac{\pi}{180}$$

$$\cos\left(\frac{\pi}{3} + h\right) \approx \frac{1}{2}\left(1 - \frac{h^2}{2}\right) - \frac{\sqrt{3}}{2}h$$

$$\cos(60.001^\circ) \approx 0.49998\,488$$

1.2.39.

$$f(x+h) = f(x) + hf'(x) + \frac{h^2}{2!}f''(x) + \frac{h^3}{3!}f'''(x) + \cdots$$

$$\begin{aligned} \text{Let} \quad f(x) &= \sin x, & f\left(\frac{\pi}{4}\right) &= \frac{1}{\sqrt{2}} \\ f'(x) &= \cos x, & f'\left(\frac{\pi}{4}\right) &= \frac{1}{\sqrt{2}} \\ f''(x) &= -\sin x, & f''\left(\frac{\pi}{4}\right) &= -\frac{1}{\sqrt{2}} \\ f'''(x) &= -\cos x, & f'''\left(\frac{\pi}{4}\right) &= \frac{1}{\sqrt{2}} \end{aligned}$$

$$\sin\left(\frac{\pi}{4} + h\right) = \frac{1}{\sqrt{2}}\left(1 + h - \frac{h^2}{2} - \frac{h^3}{3!} + \frac{h^4}{4!} + \frac{h^5}{5!} - \cdots\right)$$

$$\sin\left(\frac{\pi}{4} + h\right) = \frac{1}{\sqrt{2}}\sum_{k=0}^{\infty}(-1)^k\left[\frac{h^{2k}}{(2k)!} + \frac{h^{2k+1}}{(2k+1)!}\right]$$

$$\begin{aligned} \sin(45.0005^\circ) &= \sin\left(\frac{\pi}{4} + \frac{\pi}{36} \times 10^{-4}\right) \\ &= \frac{1}{\sqrt{2}}(1.0000087) \\ \sin(45.0005^\circ) &\approx 0.70711295 \end{aligned}$$

1.2.42. $f(x-h) = (x-h)^m = x^m - mhx^{m-1} + m(m-1)\frac{h^2}{2!}x^{m-2} + \cdots$

1.2.47.

$$\begin{aligned} & n! > \sqrt{2\pi n}\; n^n e^{-n} \\ & \frac{1}{n!} < \frac{1}{\sqrt{2\pi n}\; n^n} e^n < \frac{1}{2}10^{-14} \\ & n = 16: \quad 4.8 \times 10^{-14} > \frac{1}{2}10^{-14} \\ & n = 17: \quad 2.82 \times 10^{-15} < \frac{1}{2}10^{-14} \end{aligned}$$

1.2.50. **b.**

$$\lim_{x\to 0}\frac{\arctan x}{x} = \lim_{x\to 0}\frac{1/(1+x^2)}{1} = 1$$

c.

$$\lim_{x\to\pi}\frac{\cos x + 1}{\sin x} = \lim_{x\to\pi}\frac{-\sin x}{\cos x} = \frac{0}{-1} = 0$$

1.2.51. Suppose $x > 0$. We want each term (without regard to sign) to be less than the previous term. Hence, we have

$$\frac{x^3}{3!} < x \quad \Rightarrow x < \sqrt{6} \approx 2.4$$
$$\frac{x^5}{5!} < \frac{x^3}{3!} \Rightarrow x < \sqrt{20} \approx 4.5$$
$$\frac{x^7}{7!} < \frac{x^5}{5!} \Rightarrow x < \sqrt{42} \approx 6.5$$

Similarly, for $x < 0$.

Hence, we require $|x| < \sqrt{6}$ and the error involved in adding n terms would be less than the absolute value of the $(n+1)$-term; namely, $|x|^{2n+1}/(2n+1)!$

$$|\text{Error}| \le \frac{|x|^{2n+1}}{(2n+1)!}$$

Want

$$\frac{23^{2n+1}}{(2n+1)!} \le 10^{-8}$$
$$(2n+1)\log 23 - \log(2n+1)! \le -8$$

Therefore, we obtain $(2n+1)\log 23 + 8 \le \log(2n+1)!$

$$n = 37: \quad 110.13 \not< 109.39$$
$$n = 38: \quad 112.85 \le 113.16$$

It would require 38 terms to compute $\sin(23)$ with error at most 10^{-8}.

Now $\sin(23) = \sin(23 - 3(2\pi))$ and $23 - 3(2\pi) \approx 4.15$.

The convergence of the series would be faster.

$$(2n+1)\log(23 - 6\pi) + 8 \le \log(2n+1)!$$

$$n = 10: \quad 20.98 \not< 19.71$$
$$n = 11: \quad 22.22 \le 22.41$$

Hence, using the periodicity of sine, the number of terms required has been reduced from 38 to 11.

$$\sin x = x - \frac{x^3}{3!} + \frac{x^5}{5!} - \frac{x^7}{7!} + \cdots$$

For the i-th term, we use t_i: $t_1 = x$, $t_2 = \dfrac{x^2}{3 \cdot 2} t_1$, $t_3 = \dfrac{x^2}{5 \cdot 4} t_2$, $\vdots$ Now

$$\begin{cases} t_1 = x \\ t_{k+1} = \dfrac{x^2}{(2k+1)(2k)} t_k, & k = 1, 2, 3, \ldots \end{cases}$$

1.2.52.

$$e^{-t^2} = 1 - t^2 + \frac{t^4}{2!} - \frac{t^6}{3!} + \frac{t^8}{4!} - \frac{t^{10}}{5!} + \cdots$$

$$\int_0^x e^{-t^2}\,dt = t - \frac{t^3}{3!} + \frac{t^5}{5\cdot 2!} - \frac{t^7}{7\cdot 3!} + \frac{t^9}{9\cdot 4!} - \frac{t^{11}}{11\cdot 5!} + \cdots \Big|_0^x$$

$$= x - \frac{x^3}{3} + \frac{x^5}{5\cdot 2!} - \frac{x^7}{7\cdot 3!} + \cdots$$

Therefore, we have $\operatorname{erf}(x) = \frac{2}{\sqrt{\pi}} \int_0^x e^{-t^2}\,dt$

$$\operatorname{erf}(x) = \frac{2}{\sqrt{\pi}} \left\{ x - \frac{x^3}{3} + \frac{x^5}{5\cdot 2!} - \frac{x^7}{7\cdot 3!} + \cdots \right\}$$

$$f(x) = \frac{2}{\sqrt{\pi}} \int_0^x e^{-t^2}\,dt \quad \Rightarrow \quad f(0) = 0$$

$$f'(x) = \frac{2}{\sqrt{\pi}} e^{-x^2} \quad \Rightarrow \quad f'(0) = \frac{2}{\sqrt{\pi}}$$

$$f''(x) = \frac{2}{\sqrt{\pi}} e^{-x^2}(-2x) \quad \Rightarrow \quad f''(0) = 0$$

$$f'''(x) = \frac{2}{\sqrt{\pi}} (e^{-x^2}(4x^2) + e^{-x^2}(-2)) \quad \Rightarrow \quad f'''(0) = \frac{2}{\sqrt{\pi}}(-2)$$

$$f^{(4)}(x) = \frac{2}{\sqrt{\pi}} (e^{-x^2}(-8x^3) + e^{-x^2}(8x) + e^{-x^2}(4x))$$

$$= \frac{2}{\sqrt{\pi}} (e^{-x^2}(-8x^3 + 12x)) \quad \Rightarrow \quad f^{(4)}(0) = 0$$

$$f^{(5)}(x) = \frac{2}{\sqrt{\pi}} (e^{-x^2}(-2x)(-8x^3 + 12x)$$

$$+ e^{-x^2}(-24x^2 + 12)) \quad \Rightarrow \quad f^{(5)}(0) = \frac{2}{\sqrt{\pi}}(12)$$

and so on. We obtain

$$\operatorname{erf}(x) = \frac{2}{\sqrt{\pi}} \left\{ x - \frac{x^3}{3} + \frac{x^5}{5\cdot 2!} - \cdots \right\}$$

Yes, the two series are the same.

$$\operatorname{erf}(1) = \frac{2}{\sqrt{\pi}} \left\{ 1 - \frac{1}{3} + \frac{1}{10} - \frac{1}{42} \right\} \approx 0.8382$$

1.2.53. By Series (4), we have

$$\begin{aligned}(1+x^2)^{-1} &= 1 - x^2 + x^4 - x^6 + \cdots \\ \int \frac{dx}{1+x^2} &= \int (1 - x^2 + x^4 - x^6 + \cdots)\, dx \\ &= x - \frac{x^3}{3} + \frac{x^5}{5} - \frac{x^7}{7} + \cdots\end{aligned}$$

Therefore, $\arctan x = \sum_{k=1}^{\infty} (-1)^{k+1} x^{2k-1}/(2k-1)$

$$\arctan(1) = 1 - \frac{1}{3} + \frac{1}{5} - \frac{1}{7} + \cdots$$

Want $\frac{1}{2n+1} < \frac{1}{2}10^{-10} \Rightarrow 10^{10} < n + \frac{1}{2} < n+1$. Therefore, we obtain $n > 10^{10} - 1$. Not practical to use this series since 10^{10} terms are needed and loss of significance will occur due to roundoff errors.

1.2.54.

$$\pi = 4 - 8\Big(\frac{1}{15} + \frac{1}{63} + \frac{1}{143} + \frac{1}{255} + \cdots \frac{1}{159999}\Big)$$

The convergence of this series is *very* slow requring an extremely large number of terms!

For example, 10^8 terms gives a relative error of 10^7.

This is *not* a practical way of computing π!

1.2 Computer Exercises

1.2.C1. Since $x^2 + 10^8 x + c = 0$, we have $x = \big(-10^8 \pm \sqrt{10^6 - 4c}\big)/2$. Now

$$\pm\sqrt{10^{16} - 4c} = \begin{cases} 10^8, & \text{if } c = 1 \\ 9.9999998 \times 10^7, & \text{if } c = 10^8 \end{cases}$$

Hence, we have

	$c = 1$	$c = 10^8$
x_1	0	-1
x_2	-10^8	$-9.9999998 \times 10^7 \approx -10^8$

But $x_1 x_2 = c/a$ and $x_1 = 10^8 x_2^{-1}$. Thus, these two equations should have the same roots. The zero root is 100% wrong.

1.2.C14. g converges faster. Convergence in 5 iterations.

1.2.C16. $\lambda_{50} = 12586269025$

1.2.C17. $\alpha_{50} = 28143753123$

1.3 Floating-Point Representation

1.3.1. **c.** $-8 \times 2^{-24} = 2^3 \times 2^{-24} = -2^{-21} = (-1.0)_2 \times 2^{-21}$
$c - 127 = -21$
$c = 127 - 21 = (106)_{10} = (152)_8 = (01\,101\,010)_2$
$[101\,101\,01\,0000\,0000\,0000\,0000\,0000\,0000]_2 = [\texttt{B5000000}]_{16}$

1.3.2. **d.** $(0.03125)_{10} = (0.02)_8 = (0.000\,010)_2 = 1.0 \times 2^{-5}$
$c - 127 = -5$
$c = 127 - 5 = (122)_{10} = (172)_8 = (01\,111\,010)_2$
$[001\,111\,010\,000\,0000\,0000\,0000\,0000\,0000]_2 = [\texttt{3D000000}]_{16}$
In double precision: $[\texttt{3FA0000000000000}]_{16}$
$-(0.03125)_{10}$
$[101\,111\,010\,000\,0000\,0000\,0000\,0000\,0000]_2 = [\texttt{BD000000}]_{16}$
In double precision: $[\texttt{BFA0000000000000}]_{16}$

1.3.4. **d.** $(0.234375)_{10} = (0.17)_8 = (0.001\,111)_2 = (1.111)_2 \times 2^{-3}$
$c - 127 = -3$
$c = 127 - 3 = (124)_{10} = (174)_8 = (01\,111\,100)_2$
$[001\,111\,100\,111\,0000\,0000\,0000\,0000\,0000]_2 = [\texttt{3E700000}]_{16}$
In double precision: $[\texttt{3FCE000000000000}]_{16}$

e. 492.78125
$(492)_{10} = (754)_8 = (111\,101\,100)_2$
$(0.78125)_{10} = (0.62)_8 = (0.110\,010)_2$
$(492.78125)_{10} = (111\,101\,100.\,110\,010)_2 = (1.\,11101\,100110\,010)_2 \times 2^8$
$c - 127 = 8$
$c = 127 + 8 = (135)_{10} = (207)_8 = (10\,000\,111)_2$
$[010\,000\,111\,111\,0110\,0110\,0100\,0000\,0000]_2 = [\texttt{43F66400}]_{16}$
In double precision: $[\texttt{407ECC8000000000}]_{16}$

1.3.5. **d.** $-\infty$

1.3.8. **a.** $[\texttt{CA3F2900}]_{16} = [1100\,1010\,0011\,1111\,0010\,1001\,0000\,0000]_2$
stored exponent:$(100\,1010\,0)_2 = (224)_8 = 148$
So $148 - 127 = 21$
$-(1.011\,1111\,0010\,1001\,0000\,0000)_2 \times 2^{21} = -(1011\,1111\,0010\,1001\,0000\,00)_2$
$= -(13745100)_8 = -3.131968 \times 10^6$

d. $[\texttt{4B187ABC}]_{16} = [0100\,1011\,0001\,1000\,0111\,1010\,1011\,1100]_2$
stored exponent:$(100\,1011\,0)_2 = (226)_8 = 150$
So $150 - 127 = 23$
$(1.001\,1000\,0111\,1010\,1011\,1100)_2 \times 2^{23} = (1001\,1000\,0111\,1010\,1011\,1100)_2$
$= (46075274)_8 = 9992892 = 9.992892 \times 10^6$

g. $[\text{C553E000}]_{16} = [1100\,0101\,0101\,0011\,1110\,0000\,0000\,0000]_2$
stored exponent:$(1000\,101\,0)_2 = (212)_8 = 138$
So $138 - 127 = 11$
$-(1.101\,0011\,1110\,0000\,0000\,0000)_2 \times 2^{11} = -(1101\,0011\,1110)_2 = -(6476)_8$
$= -3390 = -3.39 \times 10^3$

1.3.11. $\pm 0.b_1b_2 \times 2^m$

c. $m = -1, 0, 1$. Non negative machine numbers:

$$0,\ \frac{1}{8},\ \frac{1}{4},\ \frac{3}{8},\ \frac{1}{2},\ \frac{3}{4},\ 1,\ \frac{3}{2}$$

All together, there are 15 machine numbers.

1.3.15. relative error $= \dfrac{|x - x'|}{|x|} = \dfrac{|x - 0|}{|x|} = 1$

1.3.17. By the previous exercise, the relative error is at most $\frac{1}{2}10^{-4}$ $(\beta = 10, n = 5)$.
We shall see that this upper bound cannot be reached. Let $x = (q + r)10^m$, where $q = (.a_1a_2 \ldots a_5)_{10}$ and $r \leq \frac{1}{2}10^{-5}$.
Then we shall round down to get the machine number $x' = q10^m$.
The relative error is $\varepsilon = (x - x')/x = r/(q + r)$.
To make this as large as possible, give q its smallest value, 10^{-1}.
By computing $d\varepsilon/dr$, we see that we now should give r its largest value, $\frac{1}{2}10^{-5}$.
Then $\varepsilon = \frac{1}{2}10^{-4}/\left(1 + \frac{1}{2}10^{-4}\right)$.
The same maximum arises if we round upward.
Note that $x = 0.10000\,5$ is one solution to the exercise.
Then $\widetilde{x} = 0.1$ and

$$|x - \widetilde{x}| = \frac{1}{2}10^{-5}$$
$$|x - \widetilde{x}|/|x| = 4.99975 \times 10^{-6} \leq \frac{1}{2}10^{-5}$$

1.3.18. Assume numbers have the form $0.b_1b_2b_3b_4b_5 \times 10^m$. So

$$\begin{aligned} 1 &= 0.10000 \quad \times 10^1 \\ x &= 0.00000\,4 \times 10^1 \quad (= 4 \times 10^{-5}, \text{ for example}) \\ \hline 1 + x &= 0.10000\,4 \times 10^1 \\ &= 0.10000 \times 10^1 \quad \text{(rounds)} \end{aligned}$$

Hence, if $|x| < 5 \times 10^{-5}$ then x will have the desired property.

1.3.19. $x = \pm(0.a_1a_2 \ldots a_na_{n+1} \ldots)_\beta \times \beta^m$ where $0 \leq a_i < \beta - 1$ with $a_1 \neq 0$.
$\text{fl}(x) = \pm(0.a_1a_2 \ldots a_n)_\beta \times \beta^m$ "chopping"

$$\text{relative error} = \frac{|x - \text{fl}(x)|}{|x|} = \frac{(0.a_{n+1} \ldots)_\beta \times \beta^{m-n}}{(0.a_1a_2 \ldots a_na_{n+1} \ldots)_\beta \times \beta^m}$$
$$\leq (0.a_{n+1} \ldots)_\beta \times \beta^{1-n} \leq \beta^{1-n}$$

since $(0.a_1a_2\ldots a_na_{n+1}\ldots)_\beta \geq (0.1)_\beta = \beta^{-1}$ and $(0.a_{n+1}\ldots)_\beta \leq 1$.

Note: $\beta = 2,\ n = 24 \Rightarrow 2^{-23}$.

1.3.21. $2^{-1} < 2^{-1} + 2^{-26} < 2^{-1} + 2^{-24}$; $(2^{-1} + 2^{-26}) - 2^{-1} = 2^{-26}$;
$(2^{-1} + 2^{-24}) - (2^{-1} + 2^{-26}) = 3 \times 2^{-26}$
relative error $= 1/(1 + 2^{25}) \approx 2^{-25}$ or $3/(1 + 2^{25}) \approx 3 \times 2^{-25}$

1.3.25. $\mathrm{fl}[\mathrm{fl}(x)\,\mathrm{fl}(y)] = [x(1 + \delta_1)y(1 + \delta_2)](1 + \delta_3)$
$= xy(1 + \delta_1)(1 + \delta_2)(1 + \delta_3) \approx xy(1 + \delta_1 + \delta_2 + \delta_3)$

Since $|\delta_i| \leq 2^{-24}$, the relative error cannot exceed 3×2^{-24}.

Here x and y are not assumed to be machine numbers.

1.3.26. The machine returns

$$\begin{aligned}
z &= \mathrm{fl}\{\mathrm{fl}(x) + \mathrm{fl}[\mathrm{fl}(y)\,\mathrm{fl}(y)]\} = \mathrm{fl}\{x(1 + \delta_1) + \mathrm{fl}[y(1 + \delta_2)y(1 + \delta_2)]\} \\
&= \{x(1 + \delta_1) + [y(1 + \delta_2)y(1 + \delta_2)](1 + \delta_3)\}(1 + \delta_4) \\
&= x(1 + \delta_1)(1 + \delta_4) + y^2(1 + \delta_2)^2(1 + \delta_3)(1 + \delta_4) \\
&\approx x(1 + \delta_1 + \delta_4) + y^2(1 + 2\delta_2 + \delta_3 + \delta_4) \\
&= (x + y^2) + [(\delta_1 + \delta_4)x + (2\delta_2 + \delta_3 + \delta_4)y^2]
\end{aligned}$$

If $x > 0$ then $x + y^2 \neq 0$ and

the relative error $\left|(x + y^2 - z)/(x + y^2)\right|$ cannot exceed 4δ where $\delta = \max_{i=1,2,3,4} \delta_i$.

For the `Marc-32`, we have $4 \times 2^{-24} = 2^{-22}$.

If $x < 0$ the relative error can be infinite.

Here is an example: Let y be the square root of 2, and let $x = -2$. Then $x + y^2 = 0$.

Hence, if there is any roundoff error, the relative error will be infinite because in computing relative error we divide the actual error by the correct answer, which in this case is 0.

There will be roundoff error in this example because y is not a machine number.

(Note that the relative error may be infinite even if $x + y^2 \neq 0$ such as when it's value is near 0.)

1.3.30. No useful bound can be given, for the true sum S_{1000} and the computed sum S^*_{1000} could satisfy these conditions: $S_{1000} = 0$; $S^*_{1000} \neq 0$.

Then $\left|\dfrac{S_{1000} - S^*_{1000}}{S_{1000}}\right| = +\infty$. If the numbers are positive, it can be shown that
$\dfrac{|S_n - S^*_n|}{S_n} \leq (1 + 2^{-24})^n - 1$.

This upper bound is approximately $n \times 2^{-24}$. If $n = 1000$, this is about 2^{-14}, since $1000 \approx 1024 = 2^{10}$.

1.3.37. Relative error for rounding $\frac{1}{2}\beta^{1-n}$.

Relative error for chopping β^{1-n}.

Since $\beta = 10$ and $n = 13$, we have

$$\begin{cases} \frac{1}{2} \times 10^{-12} \text{ rounding} \\ \quad 10^{-12} \text{ chopping} \end{cases}$$

1.3.38.

$$\begin{array}{rr} 0.5321487513 & 0.53215 \\ -\underline{0.5321304421} & -\underline{0.53213} \\ 0.0000183092 & 0.00002 \end{array}$$

$$\left| \frac{(\text{fl}(x) - \text{fl}(y)) - (x - y)}{(x - y)} \right| = \frac{2. \times 10^{-5} - 1.83092 \times 10^{-5}}{1.83092 \times 10^{-5}} = 9.235 \times 10^{-2}$$

Therefore, we obtain the relative error $\approx 9\%$.

1.3.39.

$$\begin{aligned} \text{fl}(xy^2) &= \text{fl}(x)\,\text{fl}(y^2)(1+\delta_1) = \text{fl}(x)\,\text{fl}(y)\,\text{fl}(y)(1+\delta_2)(1+\delta_1) \\ &= x(1+\delta_3)y(1+\delta_4)y(1+\delta_5)(1+\delta_2)(1+\delta_1) \\ &= xy^2(1+\delta_1)(1+\delta_2)(1+\delta_3)(1+\delta_4)(1+\delta_5) \\ &\approx xy^2(1+\delta_1+\delta_2+\delta_3+\delta_4+\delta_5) \end{aligned}$$

Since $|\delta_i| \le 2^{-24}$, the relative error cannot exceed 5×2^{-24}.

1.3.42. $((q - 2^{-25})2^m,\ (q + 2^{-25})2^m)$

1.3 Computer Exercises

1.3.C3. 1.41421 35623 73095

1.3.C7. $s \le 33.2710$

1.3.C8. Note that the two limits are equal because

$\lim_{m\to\infty}[\log(m + \frac{1}{2}) - \log m]$

$= \lim_{m\to\infty}[\log(m + \frac{1}{2})/m]$

$= \lim_{m\to\infty}\{\log[1 + 1/(2m)]\}$

$= \log 1 = 0.$

1.4 Loss of Significance

1.4.4. $y = (1 - \sin x)[(1 + \sin x)/(1 + \sin x)] = (1 - \sin^2 x)/(1 + \sin x) = \cos^2 x/(1 + \sin x)$

$$y \leftarrow \cos^2 x/(1 + \sin x)$$

1.4.6. $f(x) = (\tan x - \sin x)/(x - \sqrt{1 + x^2})$

$\tan x = x + (x^3/3) + (2/15)x^5 + \cdots$

$\underline{\sin x = x - (x^3/3!) + (x^5/5!) + \cdots}$

$\text{Numer} = (1/2)x^3 + (1/8)x^5$

$\text{Denom} = x - (1 + x^2)^{1/2}$

$= x - (1 + (1/2)x^2 - (1/8)x^4 - \cdots)$

$= -1 + x - (1/2)x^2 + (1/8)x^4 + \cdots$

$[(1/2)x^3 + (1/3)x^5 + \cdots]/[-1 + x - (1/2)x^2 + \cdots] = -(1/2)x^3 - (1/2)x^4 - (1/4)x^5 - \cdots$

$f(x) \approx -(1/2)x^3 - (1/2)x^4$

$f(0.0125) \approx -9.888 \times 10^{-7}$

1.4.8. $f(x) = g(x) - h(x)$ where

$$g(x) = \frac{\sqrt{1 + x^2} - 1}{x^2}, \qquad h(x) = \frac{x^2 \sin x}{x - \tan x}\,.$$

When $x \approx 0$, $g(x)$ involves loss of significance in the numerator and $h(x)$ in the denominator. Now

$$g(x) = \frac{(\sqrt{1 + x^2} - 1)}{x^2} \frac{(\sqrt{1 + x^2} + 1)}{(\sqrt{1 + x^2} + 1)} = \frac{1}{\sqrt{1 + x^2} + 1}$$

Moreover, we have

$$h(x) = x^2 \Big(\frac{x - (x^3/3!) + (x^5/5!) - \cdots}{x - x - (x^3/3) - (2x^5/15) - \cdots} \Big)$$

$$\approx - \Big(\frac{x - (x^3/6)}{(x/3) + (2/15)x^3} \Big) = -3 + \frac{17}{10}x^2 + \cdots$$

Therefore, we obtain $h(x) \approx -3 + (17/10)x^2$.

Hence, we obtain $f(x) = (1/\sqrt{1 + x^2} + 1) + 3 - 1.7x^2$.

Clearly, $f(0) = (1/2) + 3 = 3.5$.

1.4.10. For large x, $\sqrt{x^2+1} \approx x$ and $f(x)$ involves loss of significance.

$$f(x) = \left[\sqrt{x^2+1} - x\right]\left[\frac{\sqrt{x^2+1}+x}{\sqrt{x^2+1}+x}\right] = \frac{1}{\sqrt{x^2+1}+x}$$

1.4.11. $f(x) = \ln(x + \sqrt{x^2+1})$

$$\begin{aligned} f(-x) &= \ln[(-x+\sqrt{x^2+1})(x+\sqrt{x^2+1})/(x+\sqrt{x^2+1})] \\ &= \ln[1/(x+\sqrt{x^2+1})] \\ &= \ln(x+\sqrt{x^2+1})^{-1} \\ &= -\ln(x+\sqrt{x^2+1}) = -f(x) \end{aligned}$$

$$f(x) = \begin{cases} \ln(x+\sqrt{x^2+1}), & x > 0 \\ 0, & x = 0 \\ -\ln(-x+\sqrt{x^2+1}), & x < 0 \end{cases}$$

1.4.13. $z = \sqrt{x^4+4} - 2$ involves loss of significance when x near zero.

$$(\sqrt{x^4+4} - 2)\frac{(\sqrt{x^4+4}+2)}{(\sqrt{x^4+4}+2)} = \frac{x^4}{\sqrt{x^4+4}+2}$$

$$z \leftarrow x^4/(\sqrt{x^4+4}+2)$$

1.4.16. $f(x) = (\cos x - e^{-x})/\sin x$

$\cos x = 1 - (x^2/2) + (x^4/4!) - (x^6/6!) + \cdots$

$\underline{e^{-x} = 1 - x + (x^2/2) - (x^3/3!) + (x^4/4!) - (x^5/5!) + (x^6/6!) + \cdots}$

$\text{Num} = x - x^2 + (x^3/3!) + (x^5/5!) - (2x^6/6) + \cdots$

$\text{Den} = x - (x^3/3!) + (x^5/5!) - \cdots$

$(x - x^2 + (x^3/6) + (x^5/120) + \cdots)/(x - (x^3/6) + (x^5/120) + \cdots)$
$= 1 - x + (x^3/3) - (x^3/6) + \cdots$

$f(x) \approx 1 - x + (x^2/3) - (x^3/6)$

$f(8 \times 10^{-3}) \approx 1 - 8 \times 10^{-3} + (64/3) \times 10^{-6} - (512/6) \times 10^{-9}$

$$\begin{array}{r} 1.00002\,1000 \\ -\underline{0.00800\,0085} \\ 0.99202\,0915 \end{array}$$

1.4.20. $\arctan x - x = x - (x^3/3) + (x^5/5) - (x^7/7) + \cdots - x$
$\approx x^3(-(1/3) + x^2((1/5) + x^2(-(1/7))))$

1.4.22. $e^{2x} = 1 + 2x + ((2x)^2/2) + ((2x)^3/6) + ((2x)^4/24) + \cdots$
$(e^{2x} - 1)/2x = 1 + x + (2/3)x^2 + (1/3)x^3 + \cdots \approx 1 + x(1 + (x/3)(2 + x))$

1.4.24. **a.** If x and y are near $\pi/2$, $\sin x$ can be very near $\sin y$ since the curve is relatively flat near $\pi/2$.

1.4.26. **b.** $\ln x - 1 = \ln(x/e)$

d. $x^{-2}(\sin x - e^x + 1) = x^{-2}(x - (x^3/3!) + \cdots - 1 - x - (x^2/2) - (x^3/3!) + \cdots + 1)$
$\approx -(1/2) - (x/3)$ when $x \to 0$

1.4.28. $(\sin x/x) = (1/x)(x - (x^3/3!) + (x^5/5!) - \cdots) = 1 - (x^2/6) + (x^4/120) - \cdots$. Suppose $x^2/6 < \epsilon$. Then $x < \sqrt{6\epsilon}$ when $x > 0$ and $x > -\sqrt{6\epsilon}$ when $x < 0$. Thus, we have $|x| < \sqrt{6\epsilon}$ where ϵ is the machine precision.

1.4.29. $x^2 - 10^5x + 1 = 0$; $x = (1/2)(10^5 \pm \sqrt{10^{10} - 4}) \approx 1/2(10^5 \pm 10^5)$; $\begin{cases} x_1 = 10^5 \\ x_2 = 0 \end{cases}$

Now $x_1^2 - 10^5x_1 + 1 \approx 10^{10} - 10^{10} = 0$ but $x_2^2 - 10^5x_2 + 1 \neq 0$.

$$\left(\frac{-b - \sqrt{b^2 - 4ac}}{(2a)}\right)\left(\frac{-b + \sqrt{b^2 - 4ac}}{-b + \sqrt{b^2 - 4ac}}\right) = \frac{2c}{-b + \sqrt{b^2 - 4ac}}$$

Hence, if $b < 0$ use

$$x_1 = \frac{-b + \sqrt{b^2 - 4ac}}{2a} = \frac{1}{2}(10^5 + \sqrt{10^5 - 4}) \approx 10^5$$

and

$$x_2 = \frac{2c}{-b + \sqrt{b^2 - 4ac}} = \frac{2}{(10^5 + \sqrt{10^5 - 4})} \approx 10^{-5}$$

1.4.30. Not much. Expect to compute $b^2 - 4ac$ in double precision.

1.4 Computer Exercises

1.4.C1. No solution; 0; 0.0; Any solution; (−1.0, 0.0);
(−0.10208 42383, −4.89791 57617); (4.00000 00001, 4.00090 99999);
(−0.10208 42383, −4.89791 57617); (0.10000 00000, 1.00000 00000E+34);
(1.99683 77223, 2.00316 22777).

1.4.C10.

x	exponent (x)	n
0.	1.0	1
1.	2.71828 18285	10
−1.	0.36787 94412	10
0.5	1.64872 12707	8
−0.123	0.88426 36626	5
−25.5	8.42346 37545 $\times 10^{-12}$	25
−1776.	0.	25
3.14159	23.14063 12270	17

1.4.C14. $|x| < 10^{-15}$

1.4.C15. $\rho_{50} = 2.85987$

Chapter 2

Linear Systems

2.1 Naive Gaussian Elimination

2.1.1.

$$\begin{bmatrix} 1 & 4 & \alpha & 6 \\ 2 & -1 & 2\alpha & 3 \\ \alpha & 3 & 1 & 5 \end{bmatrix} \quad \Rightarrow \quad \begin{bmatrix} 1 & 4 & \alpha & 6 \\ 0 & -9 & 0 & -9 \\ 0 & 3-4\alpha & 1-\alpha^2 & 5-6\alpha \end{bmatrix}$$

$$x_2 = 1$$

$$\begin{cases} (1-\alpha^2)x_3 = 2(1-\alpha) \\ x_1 = 2 - \alpha x_3 \end{cases}$$

$$\begin{aligned} \alpha &= 0: && x_1 = 2,\ x_2 = 1,\ x_3 = 2 \\ \alpha &= -1: && \text{no solution} \quad (0x_3 = 4) \\ \alpha &= 1: && \text{infinite number of solutions} \quad (0x_3 = 0) \end{aligned}$$

Homogeneous:

$$x_2 = 0$$

$$\begin{cases} (1-\alpha^2)x_3 = 0 \\ x_1 = -\alpha x_3 \end{cases}$$

$$\begin{aligned} \alpha &= 0: && x_1 = x_2 = x_3 = 0 \\ \alpha &= 1: && \text{infinite number of solutions} \quad (0x_3 = 0) \\ \alpha &= -1: && \text{infinite number of solutions} \quad (0x_3 = 0) \end{aligned}$$

2.1.2.

$$\begin{bmatrix} 1 & 1 & 2 \\ \alpha & 1 & 2+\alpha \end{bmatrix} \quad \Rightarrow \quad \begin{bmatrix} 1 & 1 & 2 \\ 0 & 1-\alpha & 2-\alpha \end{bmatrix}$$

$$x_2 = (2-\alpha)/(1-\alpha), \ x_1 = -\alpha/(1-\alpha).$$

For $\alpha \approx 1$, erroneous answer is produced.

2.1.3. **a.** After forward elimination:

$$\begin{cases} 3x_1 + 2x_2 = 4 \\ 0x_2 = \dfrac{7}{3} \end{cases}$$

Multiply second equation by -3 and notice that the original equations correspond to two parallel lines in R^2. Hence, we obtain no solution.

b. After forward elimination:

$$\begin{cases} 6x_1 - 3x_2 = 6 \\ 0x_2 = 0 \end{cases}$$

Multiply second equation by -3 and notice that the original equations correspond to one line in R^2. Hence, we obtain an infinite number of solutions.

2.1.4.

$$\begin{cases} .1036x + .2122y = .7381 \\ \quad - .001610y = -.5501 \end{cases} \Rightarrow \begin{cases} y = 341.7 \\ x = -692.8 \end{cases}$$

$$\begin{cases} .1036x + .2122y = .7381 \\ \quad - .001543436y = -.54991207 \end{cases} \Rightarrow \begin{cases} y = 356.39081 \\ x = -722.65260 \end{cases}$$

2.1.5.

$$\mathbf{r} = \begin{bmatrix} -0.001343 \\ -0.001572 \end{bmatrix}, \quad \widehat{\mathbf{r}} = \begin{bmatrix} -0.0000001 \\ 0.0000000 \end{bmatrix}, \quad \mathbf{e} = \begin{bmatrix} -0.001 \\ -0.001 \end{bmatrix}, \quad \widehat{\mathbf{e}} = \begin{bmatrix} -0.659 \\ 0.913 \end{bmatrix}$$

According to the residual vectors, $\widehat{x}$ is the solution but, according to the error vectors, x is. This is an ill-conditioned problem since $\det(\mathbf{A}) \approx 10^{-6}$.

2.1.6. **a.** After forward elimination:

$$\begin{cases} 10^{-4}x_1 + x_2 = 1 \\ (1 - 10^4)x_2 = (2 - 10^4) \end{cases}$$

Thus, we have $x_2 = 1,\ x_1 = 0$.

b. After forward elimination:

$$\begin{cases} x_1 + x_2 = 2 \\ \quad (1 - 10^{-4})x_2 = (1 - 2 \times 10^{-4}) \end{cases}$$

Thus, we have $x_2 = 1,\ x_1 = 1$.

c. After forward elimination:

$$\begin{cases} 10^{-4}x_1 + x_2 = b_1 \\ (1 - 10^4)x_2 = (b_2 - b_1 10^4) \end{cases}$$

Let $b_1 = b_2 = 1$. Then $x_2 = 1,\ x_1 = 0$, which is exact.

2.1.7. **a.** Forward elimination:

$$\begin{cases} 3x_1 + 4x_2 + 3x_3 = 10 \\ \frac{11}{3}x_2 - 2x_3 = \frac{11}{3} \\ -5x_2 + x_3 = -5 \end{cases}$$

Finally, we have

$$\begin{cases} 3x_1 + 4x_2 + 3x_3 = 10 \\ \frac{11}{3}x_2 - 2x_3 = \frac{11}{3} \\ -\frac{19}{11}x_3 = 0 \end{cases}$$

Back substitution:

$$x_3 = 0,\ x_2 = 1,\ x_1 = 2$$

b. Forward elimination:

$$\begin{cases} 3x_1 + 2x_2 - 5x_3 = 0 \\ -\frac{13}{3}x_2 + \frac{13}{3}x_3 = 0 \\ \frac{10}{3}x_2 + \frac{2}{3}x_3 = 4 \end{cases}$$

$$\begin{cases} 3x_1 + 2x_2 - 5x_3 = 0 \\ -\frac{13}{3}x_2 + \frac{13}{3}x_3 = 0 \\ 4x_3 = 4 \end{cases}$$

Back substitution:

$$x_3 = 1,\ x_2 = 1,\ x_1 = 1$$

c. Forward elimination:

$$\begin{bmatrix} 1 & -1 & 2 & 1 & 1 \\ 0 & 5 & -5 & 1 & -2 \\ 0 & 0 & 9 & -\frac{23}{5} & \frac{6}{5} \\ 0 & 0 & 0 & -\frac{2}{3} & -3 \end{bmatrix}$$

Back substitution:

$$\begin{aligned} x_4 &= \frac{9}{2} = 4.5, \\ x_3 &= \frac{73}{30} \approx 2.433, \\ x_2 &= \frac{34}{30} \approx 1.133, \\ x_1 &= -\frac{217}{30} \approx -7.233 \end{aligned}$$

2.1 Computer Exercises

2.1.C6. $\mathbf{z} = (2i, i, i, i)^T, \lambda = 1 + 5i$; $\mathbf{z} = (1, 2, 1, 1)^T$, $\lambda = 2 + 6i$;
$\mathbf{z} = (-i, -i, 0, -i)^T$, $\lambda = -3 - 7i$; $\mathbf{z} = (1, 1, 1, 0)^T$, $\lambda = -4 - 8i$.

2.1.C7. A complex linear systems of equations may arise in the analyze and design of electrical circuits. The fundamental properties are

Resistor	R (ohms, W) :	$V_R = I_R R$
Capacitor	C (farads, F) :	$V_C = I_C\,[1/\,(iC\omega)]$
Inductor	L (henrys, L) :	$V_L = I_L(iL\omega)$

From the figure, we let

$$\begin{aligned}
&R_1 = 50, R_2 = 40, R_3 = 20, L = 15\\
&C_1 = 0.1, C_2 = 0.5, C3 = 0.2 \Rightarrow C_1^{-1} = 10, C_2^{-1} = 2, C_3^{-1} = 5\\
&\omega = 1, 1/i = -i
\end{aligned}$$

The equations for the loops in the circuit are

Loop 1:

$$\begin{aligned}
R_1 I_1 + (I_1 - I_2)\Big(\frac{1}{iC_1\omega}\Big) &= V_1\\
\Big(R_1 - \frac{i}{C_1\omega}\Big)I_1 + \Big(\frac{i}{C_1\omega}\Big)I_2 &= V_1\\
(50 - 10i)I_1 + (10i)I_2 &= V_1
\end{aligned}$$

Loop 2:

$$\begin{aligned}
R_2 I_2 + \Big(\frac{1}{iC_2\omega}\Big)I_2 + (I_2 - I_1)\Big(\frac{1}{iC_1\omega}\Big) + (I_2 - I_3)(iL\omega) &= 0\\
\Big(\frac{i}{C_1\omega}\Big)I_1 + \Big(R_2 - \frac{i}{C_1\omega} - \frac{i}{C_2\omega} + iL\omega\Big)I_2 - (iL\omega)I_3 &= 0\\
(10i)I_1 + (40 + 3i)I_2 - (15i)I_3 &= 0
\end{aligned}$$

Loop 3:

$$\begin{aligned}
\Big(\frac{1}{iC_3\omega}\Big)I_3 + I_3 R_3 + (I_3 - I_2)(iL\omega) &= -V_2\\
-(iL\omega)I_2 + \Big(R_3 - \frac{i}{C_3\omega} + iL\omega\Big)I_3 &= -V2\\
-(15i)I_2 + (20 + 10i)I_3 &= -V_2
\end{aligned}$$

The complex linear system can be written as an augmented matrix

$$\left[\begin{array}{ccc|c}
50 - 10i & 10i & 0 & V_1\\
10i & 40 + 3i & -15i & 0\\
0 & -15i & 20 + 10i & -V_2
\end{array}\right]$$

a. Let $V_1 = V_2 = 100$:
The complex numerical solution is

$$\left[1.6034 + 0.4165i, \quad -0.4793 - 1.5664i, \quad -3.2039 + 1.2425i\right]^T$$

b. Let $V_1 = 100$, $V_2 = 100i$:
The complex numerical solution is

$$\left[1.7915 + 0.1034i, \quad 1.2743 - 0.9389i, \quad -1.0544 - 3.517i\right]^T$$

2.2 Gaussian Elimination with Scaled Partial Pivoting

2.2.1.

$$\begin{bmatrix} \frac{1}{2} & \frac{5}{2} & -4 & -1 \\ \frac{1}{4} & -\frac{5}{4} & 0 & -3 \\ \frac{3}{4} & \frac{9}{4} & 4 & 0 \\ 4 & 1 & 0 & 4 \end{bmatrix} \Rightarrow \begin{bmatrix} \frac{1}{2} & \frac{5}{2} & -4 & -1 \\ \frac{1}{4} & -\frac{1}{2} & -2 & -\frac{7}{2} \\ \frac{3}{4} & \frac{9}{10} & \frac{38}{5} & \frac{9}{10} \\ 4 & 1 & 0 & 4 \end{bmatrix} \Rightarrow \begin{bmatrix} \frac{1}{2} & \frac{5}{2} & -4 & -1 \\ \frac{1}{4} & -\frac{1}{2} & -\frac{5}{19} & -\frac{62}{19} \\ \frac{3}{4} & \frac{9}{10} & \frac{38}{5} & \frac{9}{10} \\ 4 & 1 & 0 & 4 \end{bmatrix}$$

2.2.2.

$$\begin{bmatrix} 1 & -1 & 2 & -2 \\ -2 & 1 & -1 & 2 \\ 4 & -1 & 2 & -1 \end{bmatrix} \Rightarrow \begin{bmatrix} 0 & -\frac{1}{2} & \frac{3}{2} & -1 \\ -2 & 1 & -1 & 2 \\ 0 & 1 & 0 & 3 \end{bmatrix} \Rightarrow \begin{bmatrix} 0 & -\frac{1}{2} & \frac{3}{2} & -1 \\ -2 & 1 & -1 & 2 \\ 0 & 0 & 3 & 1 \end{bmatrix}$$

$$\mathbf{x} = \left(\frac{1}{3}, 3, \frac{1}{3}\right)^T$$

2.2.3.

$$\begin{bmatrix} 1 & 0 & 3 & 0 \\ 0 & 1 & 3 & -1 \\ 3 & -3 & 0 & 6 \\ 0 & 2 & 4 & -6 \end{bmatrix} \Rightarrow \begin{bmatrix} 0 & 1 & 3 & -2 \\ 0 & 1 & 3 & -1 \\ 3 & -3 & 0 & 6 \\ 0 & 2 & 4 & -6 \end{bmatrix} \Rightarrow \begin{bmatrix} 0 & 1 & 3 & -2 \\ 0 & 0 & 0 & 1 \\ 3 & -3 & 0 & 6 \\ 0 & 0 & -2 & -2 \end{bmatrix}$$

2.2.5.

$$\begin{bmatrix} \frac{1}{4} & \frac{5}{2} & \frac{7}{4} & \frac{1}{2} \\ 4 & 2 & 1 & 2 \\ \frac{1}{2} & 0 & \frac{3}{2} & 2 \\ \frac{1}{4} & \frac{3}{2} & \frac{15}{4} & \frac{1}{2} \end{bmatrix} \Rightarrow \begin{bmatrix} \frac{1}{4} & \frac{5}{2} & \frac{7}{4} & \frac{1}{2} \\ 4 & 2 & 1 & 2 \\ \frac{1}{2} & 0 & \frac{3}{2} & 2 \\ \frac{1}{4} & \frac{3}{5} & \frac{27}{10} & \frac{1}{5} \end{bmatrix} \Rightarrow \begin{bmatrix} \frac{1}{4} & \frac{5}{2} & \frac{7}{4} & \frac{1}{2} \\ 4 & 2 & 1 & 2 \\ \frac{1}{2} & 0 & \frac{5}{9} & \frac{17}{9} \\ \frac{1}{4} & \frac{3}{5} & \frac{27}{10} & \frac{1}{5} \end{bmatrix}$$

2.2.6. $\mathbf{s} = [7, 3, 4], \boldsymbol{\ell} = (1, 2, 3)$

$$\begin{bmatrix} 4 & 7 & 3 \\ 0 & \frac{5}{4} & \frac{5}{4} \\ 0 & -\frac{15}{2} & -\frac{5}{2} \end{bmatrix}$$

$$\boldsymbol{\ell} = (1, 2, 3); \quad \left\{ \frac{5}{4 \times 3}, \frac{5}{2 \times 4} \right\} \Rightarrow \boldsymbol{\ell} = (1, 3, 2)$$

The second pivot row is the 3rd row.

2.2.8.

$$\begin{bmatrix} 2 & 4 & -2 & 6 \\ 1 & 3 & 4 & -1 \\ 5 & 2 & 0 & 2 \end{bmatrix} \qquad \begin{aligned} \mathbf{s} &= (4, 4, 5) \\ \boldsymbol{\ell} &= (3, 2, 1) \\ \text{Ratios} &= \left\{ \frac{2}{4}, \frac{1}{4}, \frac{5}{5} \right\} \end{aligned}$$

$$\begin{bmatrix} 0 & \frac{16}{5} & -2 & \frac{26}{5} \\ 0 & \frac{13}{5} & 4 & -\frac{7}{5} \\ 5 & 2 & 0 & 2 \end{bmatrix} \qquad \text{Ratios} = \left\{ \frac{13}{5 \times 4}, \frac{16}{5 \times 4} \right\} \Rightarrow \boldsymbol{\ell}_2 = (3, 1, 2)$$

$$\begin{bmatrix} 0 & \frac{16}{5} & -2 & \frac{26}{5} \\ 0 & 0 & \frac{45}{8} & -\frac{45}{8} \\ 5 & 2 & 0 & 2 \end{bmatrix} \qquad \Rightarrow x_3 = -1, \ x_2 = 1, \ x_1 = 0$$

2.2.10.

$$\begin{cases} -x_1 + x_2 - 3x_4 = 4 \\ x_1 + 3x_3 + x_4 = 0 \\ x_2 - x_3 - x_4 = 3 \\ 3x_1 + x_3 + 2x_4 = 1 \end{cases}$$

$$\mathbf{s} = (3,3,1,3) \quad \boldsymbol{\ell} = (1,2,3,4)$$

$$\text{Ratios} = \left\{\frac{1}{3}, \frac{1}{3}, 0, \frac{3}{3}\right\} \Rightarrow \text{row } 4 \Rightarrow \boldsymbol{\ell} = (4,2,3,1)$$

$$\begin{cases} x_2 + \frac{1}{3}x_3 - 2\frac{1}{3}x_4 = 4\frac{1}{3} \\ 2\frac{2}{3}x_3 + \frac{1}{3}x_4 = -\frac{1}{3} \\ x_2 - x_3 - x_4 = 3 \\ 3x_1 + x_3 + 2x_4 = 1 \end{cases}$$

$$\text{Ratios} = \left\{0, 1, \frac{1}{3}\right\} \Rightarrow \boldsymbol{\ell} = (4,3,2,1)$$

$$\begin{cases} 1\frac{1}{3}x_3 - 1\frac{1}{3}x_4 = 1\frac{1}{3} \\ 2\frac{2}{3}x_3 + \frac{1}{3}x_4 = -\frac{1}{3} \\ x_2 - x_3 - x_4 = 3 \\ 3x_1 + x_3 + 2x_4 = 1 \end{cases}$$

$$\text{Ratios} = \left\{\frac{2\frac{2}{3}}{3}, \frac{1\frac{1}{3}}{3}\right\} = \left\{\frac{8}{7}, \frac{4}{7}\right\} \Rightarrow \boldsymbol{\ell} = (4,3,2,1)$$

$$\begin{cases} -\frac{3}{2}x_4 = \frac{3}{2} \\ 2\frac{2}{3}x_3 + \frac{1}{3}x_4 = -\frac{1}{3} \\ x_2 - x_3 - x_4 = 3 \\ 3x_1 + x_3 + 2x_4 = 1 \end{cases}$$

$$x_4 = -1, \; x_3 = 0, \; x_2 = 2, \; x_1 = 1$$

2.2.13. b.

$$\boldsymbol{\ell} = (1,2,3),\ \mathbf{s} = (5,3,4);\ \text{Ratios} = \left\{\frac{3}{5}, \frac{2}{3}, \frac{1}{4}\right\}$$

$$\begin{cases} \quad\ \ \frac{13}{2}x_2 - \frac{13}{2}x_3 = 0 \\ 2x_1 - 3x_2 + \quad x_3 = 0 \\ \quad\ \ \frac{11}{2}x_2 - \frac{3}{2}x_3 = 4 \end{cases}$$

$$\boldsymbol{\ell} = (2,1,3),\ \text{Ratios} = \left\{\frac{13}{10}, \frac{11}{8}\right\}$$

$$\Rightarrow \begin{cases} \qquad\qquad -\frac{52}{11}x_3 = -\frac{52}{11} \\ 2x_1 - 3x_2 + \quad x_3 = \quad 0 \\ \quad\ \ \frac{11}{2}x_2 - \frac{3}{2}x_3 = \quad 4 \end{cases}$$

$$\boldsymbol{\ell} = (2,3,1)$$

Thus, we have $x_3 = 1,\ x_2 = 1,\ x_1 = 1.$

d.

$$\boldsymbol{\ell} = (1,2,3),\ \mathbf{s} = (3,5,3);\ \text{Ratios} = \left\{\frac{3}{3}, \frac{5}{5}, \frac{1}{3}\right\}$$

$$\begin{bmatrix} 3 & 2 & -1 & 7 \\ 0 & -\frac{1}{3} & \frac{11}{3} & -\frac{23}{3} \\ 0 & \frac{5}{3} & -\frac{10}{3} & \frac{4}{3} \end{bmatrix}$$

$$\boldsymbol{\ell} = (1,2,3),\ \text{Ratios} = \left\{\frac{1}{15}, \frac{5}{9}\right\}$$

$$\Rightarrow \begin{bmatrix} 3 & 2 & -1 & 7 \\ 0 & 0 & 3 & -\frac{111}{15} \\ 0 & \frac{5}{3} & -\frac{10}{3} & \frac{4}{3} \end{bmatrix}$$

$$\boldsymbol{\ell} = (1,3,2),$$

Thus, we have

$$\begin{cases} x_1 = \frac{64}{15} \approx 4.267 \\ x_2 = -\frac{62}{15} \approx -4.133 \\ x_3 = -\frac{37}{15} \approx -2.467 \end{cases}$$

2.2.17.

I	1	2	3	$\cdots$	n	Total
+	1	2	3	$\cdots$	n	$(n/2)(n+1)$
*	1	2	3	$\cdots$	n	$(n/2)(n+1)$

$\Rightarrow$ # operations $= n(n+1)$

2.2.18. For the first elimination step, a total $n-1$ rows are to be processed requiring $(2n-1)$ divisions, $(n-1)^2$ multiplications, and $(n-1)^2$ subtractions.

Consider subsequent steps we have for divisions $(2n-1)+(2n-3)+\cdots+(3) = n^2-1$, for multiplications $(n-1)^2+(n-2)^2+\cdots+1 = \frac{n}{6}(n-1)(2n-1)$, and for subtractions $\frac{n}{6}(n-1)(2n-1)$. The function of time is

$$\left[\frac{29}{10}(n^2-1) + \frac{7}{30}n(n-1)(2n-1)\right] 10^{-6} \text{ seconds.}$$

2.2.19. For $\mathbf{A}$ $n \times n$ matrix, we have

$$\frac{n^3}{3}\text{operations,} \qquad \left(\frac{n^3}{3}\right)(10^{-6})\left(\frac{1}{3600}\right)\text{hours} \approx 9.26 \times 10^{-11} \times n^3\text{hours}$$

n	10	10^2	10^3	10^4
Time	$\frac{1}{3} \times 10^{-3}$ sec.	$\frac{1}{3}$ sec.	5.56 min.	3.86 days
Cost	.005¢	5¢	\$46.30	\$46,296.30

2.2.21. After Gauss, $\mathbf{A} = \mathbf{LU}$ so $\mathbf{A}^T = \mathbf{U}^T\mathbf{L}^T$. The equation $\mathbf{A}^T\mathbf{x} = \mathbf{b}$ becomes $\mathbf{U}^T\mathbf{L}^T\mathbf{x} = \mathbf{b}$. So solve

$$\begin{cases} \mathbf{U}^T\mathbf{y} = \mathbf{b} \\ \mathbf{L}^T\mathbf{x} = \mathbf{y} \end{cases}$$

2.2.23. a.

$$\begin{bmatrix} 0 & -1 & 2\times 10^9 & 0 \\ 0 & -5 & 10^9 & -1 \\ 1 & 2 & 0 & 1 \end{bmatrix} \Rightarrow \begin{bmatrix} 0 & 0 & \frac{9}{5}\times 10^9 & \frac{1}{5} \\ 0 & -5 & 10^9 & -1 \\ 1 & 2 & 0 & 1 \end{bmatrix}$$

Hence, we obtain $x_3 = \frac{1}{9} \times 10^{-9}$, $x_2 = \frac{2}{9}$, $x_1 = \frac{5}{9}$.

2.2 Computer Exercises

2.2.C2. $(3.4606, 1.5610, -2.9342, -0.4301)^T$

2.2.C3. $(6.7831, 3.5914, -6.4451, -1.5179)^T$

2.2.C4. For $2 \le n \le 10$, computed solution such that $x_i \approx 1$ for all i. For larger n, many $x_i \neq 1$

2.2.C5. $b_i = n^2 + 2(i-1)$

2.2.C6. $x_2 = 1, x_i = 0$ for $i \neq 2$

2.3 Tridiagonal and Banded Systems

2.3.2. **a.** $3(n-1)$ multiplications, $2(n-1)+1$ divisions; number of arithmetic operations $5n-4$.

2.3.3. Storage locations: $n + 2(n-1) + 2(n-2) + \cdots + 2(n-k)$

$$= n + 2[kn - (1 + 2 + \cdots + k)] = n + 2nk - 2(k)(k+1)/2 = n + 2nk - k(k+1)$$

2.3.6. Yes, it does. First step produces

$$\begin{bmatrix} a_{11} & a_{12} & a_{13} & a_{14} & \cdots \\ 0 & a'_{22} & a'_{23} & a'_{24} & \cdots \\ & \vdots & \vdots & \vdots & \ddots \end{bmatrix}$$

where $a'_{2j} = a_{2j} - m a_{1j}$; $\quad m = a_{21}/a_{11}$.

Now

$$\begin{aligned} |a'_{22}| &= |a_{22} - m a_{12}| \ge |a_{22}| - |m||a_{12}| \\ &> \sum_{\substack{j=1 \\ j\neq 2}}^{n} |a_{2j}| - |m||a_{12}| = \sum_{j=3}^{n} |a_{2j}| + |a_{21}| - |m||a_{12}| \\ &= \sum_{j=3}^{n} |a_{2j}| + |m|\left\{|a_{11}| - |a_{12}|\right\} > \sum_{j=3}^{n} |a_{2j}| + |m|\left\{\sum_{j=3}^{n} |a_{1j}|\right\} \\ &\ge \sum_{j=3}^{n} \left\{|a_{2j}| + |m||a_{1j}|\right\} \ge \sum_{j=3}^{n} |a_{2j} - m a_{1j}| = \sum_{j=3}^{n} |a'_{2j}| \end{aligned}$$

2.3.7. Let $\mathbf{D} = [\alpha_i]$. Then $\mathbf{D}^{-1} = [\alpha_i^{-1}]$. Now

$$\mathbf{D}^{-1}\mathbf{A}\mathbf{D} = [\alpha_i^{-1}][a_{i-1}, d_i, c_i][\alpha_i] = [\alpha_i^{-1} a_{i-1}\alpha_{i-1}, d_i, \alpha_i^{-1} c_i \alpha_{i+1}]$$

Want

$$a_i(\alpha_i/\alpha_{i+1}) = c_i(\alpha_{i+1}/\alpha_i) \text{ or } (a_i/c_i) = (\alpha_{i+1}/\alpha_i)^2$$

Hence, we obtain

$$a_{i+1} = \alpha_i\sqrt{a_i/c_i} \quad (1 \le i \le n-1)$$

with α_1 arbitrary ($\alpha_i \neq 0$). Require $c_i \neq 0, a_i \neq 0, sign(a_i) = sign(c_i)$. Thus, we have

$$\begin{aligned}\mathbf{D}^{-1}\mathbf{A}\mathbf{D} &= [a_{i-1}(c_{i-1}/a_{i-1})^{\frac{1}{2}}, d_i, c_i(a_i/c_i)^{\frac{1}{2}}] \\ &= [\sqrt{a_{i-1}c_{i-1}}, d_i, \sqrt{a_i c_i}\ ] \text{ and } a_i c_i > 0 \quad (1 \le i \le n-1)\end{aligned}$$

2.3 Computer Exercises

2.3.C3. $\begin{cases} d_i \leftarrow d_i - 1/d_{i-1} \\ b_i \leftarrow b_i - b_{i-1}/d_{i-1}, \quad i = 2, \ldots, n \end{cases}$

$$\begin{cases} d_n \leftarrow b_n \\ d_i \leftarrow (b_i - d_{i+1})/d_i, \quad i = n-1, \ldots, 1 \end{cases}$$

2.3.C4. $\begin{cases} d_1 = 1 \\ d_i = 1 - (4d_{i-1})^{-1}, \quad 2 \le i \le 100 \end{cases}$

2.3.C11. **a.** $\begin{cases} x_1 \leftarrow b_1/a_{11} \\ x_i \leftarrow \left(b_i - \sum_{j=1}^{n-1} a_{ij}x_j\right)\Big/a_{ii}, \quad i = 2, 3, \ldots, n \end{cases}$

2.3.C12. $\begin{cases} c_i \leftarrow c_i/d_i \\ b_i \leftarrow b_i/d_i \\ d_{i+1} \leftarrow d_{i+1} - a_{i+1}c_i \\ b_{i+1} \leftarrow b_{i+1} - a_{i+1}b_i, \quad i = 1, 2, \ldots, n-1 \end{cases}$

$$\begin{cases} b_n \leftarrow b_n/d_n \\ b_i \leftarrow b_i - c_i b_{i+1}, \quad i = n-1, \ldots, 1 \end{cases}$$

Chapter 3

Nonlinear Equations

3.1 Bisection Method

3.1.1.

x	$e^x - 3x$
0.	1.0
1.	-0.3
0.5	0.2
0.75	-0.1
0.625	-0.007
0.5625	0.07
0.59375	0.03
0.609375	0.01
0.617188	0.002
0.621094	-0.002
0.61914	-0.00009

Roots: 0.61906, 1.51213

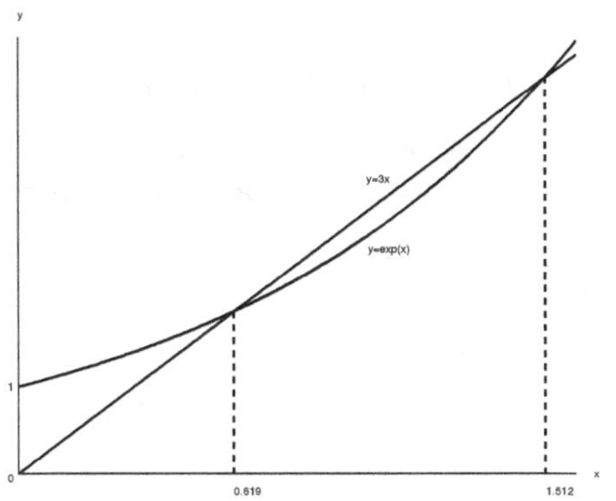

3.1.4. Approximate roots $\approx -\pi/4 - \delta, 0, \pi/4 + \varepsilon, 3\pi/4 + \varepsilon, 5\pi/4 + \varepsilon, \ldots$, where $\delta \approx 0.2$ and ε starts at approximately 0.4 and decreases.

Roots $-0.95, 0, 1.23, 2.68, 4.12, \ldots$.

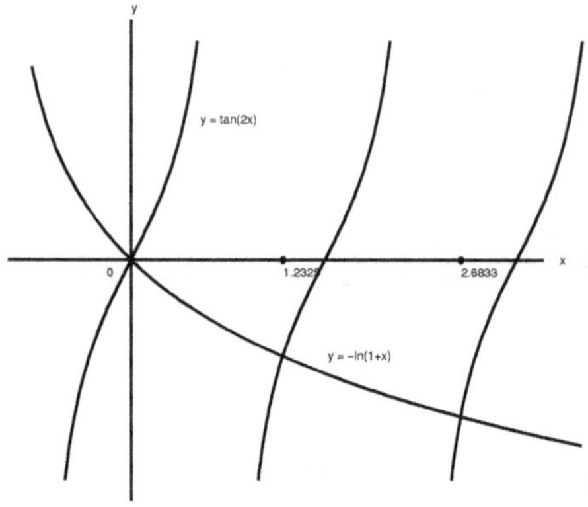

3.1.9. $\cos x - \cos 3x = 2\sin\frac{1}{2}(4x)\sin\frac{1}{2}(2x) = 2\sin 2x \sin x$.

Now $\sin x = 0$ if $x = \pm n\pi$, $n = 0, 1, 2, \ldots$; $\sin 2x = 0$ if $x = \pm n\frac{\pi}{2}$, $n = 0, 1, 2, \ldots$.

Hence, roots of $f(x)$ are $\left\{0, \pm\frac{\pi}{2}, \pm\pi, \pm\frac{3\pi}{2}, \pm 2\pi, \ldots\right\}$

3.1.10. $\left(\frac{1+x}{1-x^2}\right) = \left(\frac{(1+x)}{(1+x)(1-x)}\right) = \frac{1}{1-x}$.

When $x = 0$, $\left(\frac{1+x}{1-x^2}\right) = 1$ and $\ln\left(\frac{1+x}{1-x^2}\right) = 0$.

This is the only root.

3.1.12. Let the initial interval be $(0, 1)$.

The first estimate is the mid-point $x_1 = \frac{1}{2} = (0.100\ldots)_2$.

Since this may be in error by $1/2$, there is only one correct bit, namely the leading 0.

The next estimate will be in error by at most $1/4$, so, for example, if $x_2 = 0.01$ then the root is in $(0, 1/2)$ and consequently the correct bits are 0.0 (2 correct) and so on.

If the original interval is of width h, then after, say, k steps we have reduced the interval containing the root to width $h2^{-k}$.

From then on, we add one bit at each step.

About 3 steps are needed for each decimal digit (actually, $\log_2 10 \equiv 3.3219$).

3.1.17. After 10 steps, we have, for example, where $\widehat{x} = \frac{1}{2}(a_{10} + b_{10})$ is the approximate root.

The error is $|\frac{1}{2}(a_9 + b_9) - \frac{1}{2}(a_{10} + b_{10})| = |a_{10} - \widehat{x}|$;
$\frac{1}{2}|b_{10} - a_{10}| = \frac{1}{2}(2^{-10})|b_0 - a_0| = 2^{-11}\ |b_0 - a_0|$.

Now after n steps, we have $|$ error $| = 2^{-(n+1)}|b_0 - a_0|$.

We require $2^{-(n+1)} < \frac{1}{2}10^{-6}$. Thus, we have $-n \log 2 < -6$ and $n > \frac{6}{\log 2} = 19.9$.

So 20 steps are required.

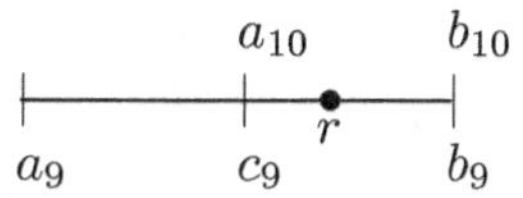

3.1.18. Bisection method $[a_0, b_0], [a_1, b_1], \ldots$

b. $|r - a_n| \le 2^{-n-1}(b_0 - a_0)$. This could be false because if r is close to b_n then $r - a_n \approx b_n - a_n = 2^{-n}(b_0 - a_0)$.

d. $0 \le r - a_n \le 2^{-n}(b_0 - a_0)$. This is true because $0 \le r - a_n$ (obvious) and $r - a_n \le b_n - a_n = 2^{-n}(b_0 - a_0)$.

3.1.19. **a.** $|r - c_n| < |r - c_{n-1}|$ false in some cases:

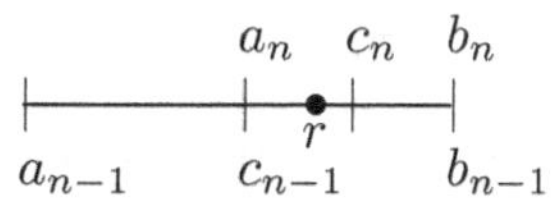

e. $|r - b_n| \le 2^{-n}(b_0 - a_0)$ true.
In fact $|r - b_n| \le |b_n - a_n| = 2^{-n}(b_0 - a_0)$.

3.1.21. By (2), $n \ge -\log 2\varepsilon / \log 2$ since unit interval.
Now $\varepsilon = 2^{-24+m}$ so $\log 2\varepsilon = (-24 + m)\log 2$ and $n \ge 24 - m$.

3.1.23. No. The real polynomial $x^2 + 1$ has only complex roots $+i, -i$.

No. The function $e^x = 1 + x + \frac{x^2}{2!} + \frac{x^3}{3!} + \cdots$ has no zeros.

3.1 Computer Exercises

3.1.C10. $1, 2, 3, 3 - 2i, 3 + 2i, 5 + 5i, 5 - 5i, 16$

3.1.C11. 2.365

3.2 Newton's Method

3.2.3. Solve $f(x) = x^2 - 1/R$, by Newton's method.

Iteration is $x_{n+1} = \dfrac{1}{2}\left[x_n + \dfrac{1}{Rx_n}\right]$.

Then $x_0 = 1$, $x_1 = 0.6$, $x_2 = 0.4666$, $x_3 = 0.4476, \ldots$ and

$x_0 = -1$, $x_1 = -3/5 = -0.6$, $x_2 = -7/15 = -0.4666$, $x_3 = -47/105 = -0.4476, \ldots$

Wrong Alternative Solution: $f(x) = x^{-2} - R$ has roots $x = 1/\pm\sqrt{R}$.

$f'(x) = -2x^{-3}$.

So we have $x - \dfrac{f}{f'} = x - \dfrac{x^{-2} - R}{-2x^{-3}} = \dfrac{-3x^{-2} + R}{-2x^{-3}} = \dfrac{1}{2}(3x - Rx^3)$.

Therefore, we have $x_{n+1} = \dfrac{x_n}{2}(3 - Rx_n^2)$. $R = 5$: $x_0 = 1$, $x_1 = -1$, $x_2 = 1$; $x_0 = -1$, $x_1 = 1$, $x_2 = -1$.

3.2.4. $g(x) = x^4 + 2x^3 - 7x^2 + 3 = 3 - x^2(7 - x(2 + x))$,
$g'(x) = 4x^3 + 6x^2 - 14x = -2x(7 - x(3 + 2x))$.

x	0	1	2
$g(x)$	3	−1	7

There is a root in the interval $[0, 1]$ and in $[1, 2]$. $x_0 = 1$, $x_1 = 0.75$, $x_2 = 0.791$, $x_3 = 0.791$. $x_0 = 2$, $x_1 = 1.75$, $x_2 = 1.64$, $x_3 = 1.62$, $x_4 = 1.62$.

Positive roots 0.79, 1.6.

3.2.7. $y = f'(x_0)(x - x_0) + f(x_0)$ where $f(x) = \sin x$, $x_0 = \dfrac{\pi}{4}$.

So we have $y = \cos\left(\dfrac{\pi}{4}\right)\left(x - \dfrac{\pi}{4}\right) + \sin\dfrac{\pi}{4} = \dfrac{1}{\sqrt{2}}\left(x - \dfrac{\pi}{4}\right) + \dfrac{1}{\sqrt{2}}$.

Therefore, we obtain $y = \dfrac{\sqrt{2}}{2}x + \dfrac{\sqrt{2}}{2}\left(1 - \dfrac{\pi}{4}\right)$.

When $y = 0, x = \dfrac{\pi}{4} - 1$ which is x_1 for Newton's method starting with $x_0 = \dfrac{\pi}{4}$.

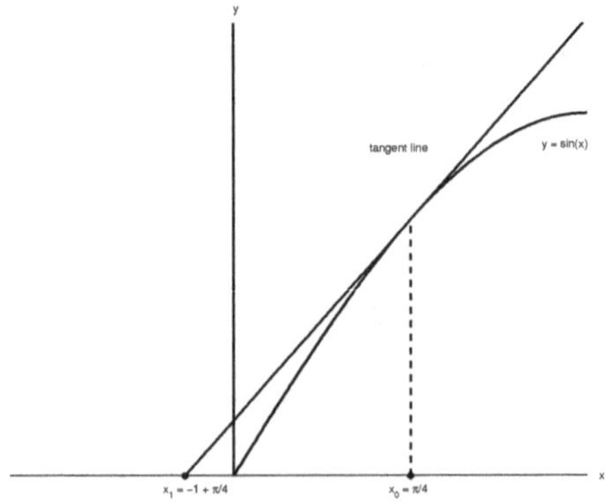

3.2.9. $x_{n+1} = x_n - \tan x_n = x_n - \dfrac{\sin x_n}{\cos x_n}$, $f(x) = \sin x$. Clearly, $\lim_{n\to\infty} x_n = \pi$ since $x_0 = 3$.

3.2.11. $f(x) = xR - x^{-1}$, $f'(x) = R + x^{-2}$.

So we have $x - \dfrac{f(x)}{f'(x)} = x - \dfrac{xR - x^{-1}}{R + x^{-2}} = x - \dfrac{x^3 R - x}{x^2 R + 1} = \dfrac{2x}{x^2 R + 1}$.

Hence, we obtain $x_{n+1} = 2x_n/(x_n^2 R + 1)$ $xR - x^{-1} = 0 \Rightarrow x = 1/\pm\sqrt{R}$.

Let $R = 4$ and $x_0 = -1$.

Then $x_1 = -2/5 = -0.4$; $x_2 = -0.4878$; $x_3 = -0.49985$

3.2.12. **a.** $x_{n+1} = \dfrac{1}{3}\left(2x_n - \dfrac{R}{x_n^2}\right) = \dfrac{2x_n^3 - R}{3x_n^2} = x_n - x_n + \dfrac{2x_n^3 - R}{3x_n^2}$

$= x_n - \dfrac{-x_n^3 - R}{3x_n^2} = x_n - \dfrac{x_n^3 + R}{3x^2}$ $f(x) = x^3 + R = 0 \Rightarrow x = -\sqrt[3]{R}$

3.2.13. **a.** $x_{n+1} = \dfrac{1}{3}(2x_n + R/x_n^2)$;

c. $x_{n+1} = x_n(x_n^3 + 2R)/(2x_n^3 + R)$;

e. $x_{n+1} = \dfrac{x_n}{3R}(4R - x_n^3)$;

g. $x_{n+1} = \dfrac{R}{x_n^2}(2x_n^6 + 1)/(2Rx_n^3 + 1)$;

3.2.15. $f(x) = x^3 - x + 1$, $f(1) = 1$; $f'(x) = 3x^2 - 1$, $f'(1) = 2$. So we obtain

$x_1 = x_0 - \dfrac{f(x_0)}{f'(x_0)} = 1 - \dfrac{1}{2} = \dfrac{1}{2}$.

3.2.17. $f(x) = x^5 - x^3 + 3$, $f(1) = 3$; $f'(x) = 5x^4 - 3x^2$, $f'(1) = 2$. So we obtain

$x_{n+1} = x_n - \dfrac{f(x_n)}{f'(x_n)} = 1 - \dfrac{3}{2} = -\dfrac{1}{2}$

3.2.19. Let $f(x) = x^2(1+^2)^{-1}$, and $f'(x) = 2x(1+x^2)^{-1} - (1+x^2)^{-2}(2x)x^2$
$= (1+x^2)^{-2}(2x + 2x^3 - 2x^3) = 2x(1+x^2)^{-2}$

So we obtain $x - \dfrac{f}{f'} = x - \dfrac{x^2(1+x^2)^{-1}}{2x(1+x^2)^{-2}} = x - \dfrac{x}{2}(1+x^2) = \dfrac{x}{2} - \dfrac{x^3}{2}.$

Therefore, we have $x_{n+1} = \dfrac{x_n}{2}(1 - x_n^2).$

From the graph of $f(x)$, we see that Newton's method converges to the only root 0 whenever $|x_n| \le 1$. But when $|x_n| \ge 2$, the method diverges.

To find the starting values x_0 for convergence, we note that convergence depends on $|x_{n+1}| < |x_n|$, i.e., $|x_n(1 - x_n^2)/2| < |x_n|$ or $|1 - x_n^2| < 2$.

For this we need $|x_0| < \sqrt{3}$.

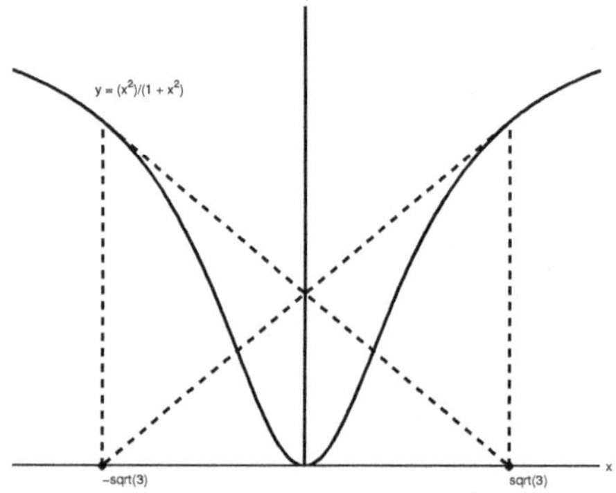

3.2.21.

$$f(x) = \begin{cases} x^{\frac{1}{2}}, & x \ge 0 \\ (-x)^{\frac{1}{2}}, & x < 0 \end{cases} \qquad f'(x) = \begin{cases} \frac{1}{2}x^{-\frac{1}{2}}, & x \ge 0 \\ -\frac{1}{2}(-x)^{-\frac{1}{2}}, & x < 0 \end{cases}$$

$$x - f(x)/f'(x) = \begin{cases} -x, & x \ge 0 \\ -x, & x < 0 \end{cases}$$

Hence, we obtain $x_{n+1} = -x_n$ and Newton's method cycles if $x_0 \ne 0$.

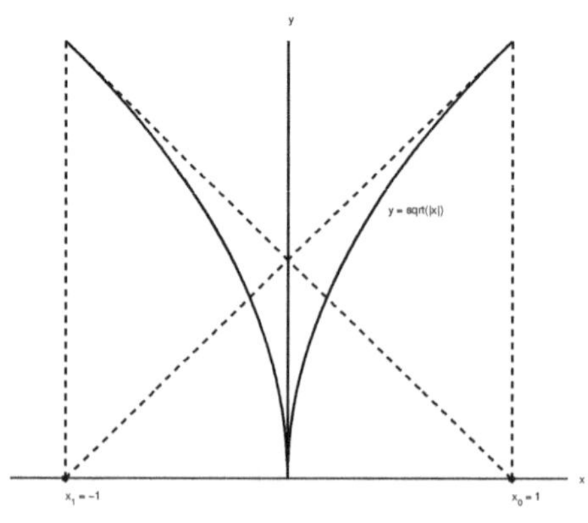

3.2.22. $f(x) = x^3 - R$, $f'(x) = 3x^2$.

So we have $x - f(x)/f'(x) = x - (x^3 - R)/(3x^2) = (2x^3 + R)/(3x^2)$.

Therefore, we obtain

$$x_{n+1} = \frac{1}{3}(2x_n + R/x_n^2)$$

$x \leftarrow R$
for $n = 1$ **to** 4
  $x \leftarrow (2x + R/x^2)/3$
end for

3.2.27. $f(x) = x^m - R$, $f'(x) = mx^{m-1}$: $x - f(x)/f'(x) = x - (x^m - R)/(mx^{m-1})$
$= [(m-1)x^m + R]/(mx^{m-1})$.

Therefore, we have $x_{n+1} = [(m-1)x_n^m + R]/(mx_n^{m-1})$.

So we have $f(x) = 1 - (R/x^m) = 1 - Rx^{-m}$, $f'(x) = mRx^{-m-1}$ and $x - f(x)/f'(x)$
$= x - (1 - Rx^{-m})/(mRx^{-m-1}) = [(m+1)Rx^{-m} - 1]/(mRx^{-m-1})$.

Therefore, we obtain $x_{n+1} = x_n[(m+1)R - x_n^m]/(mR)$.

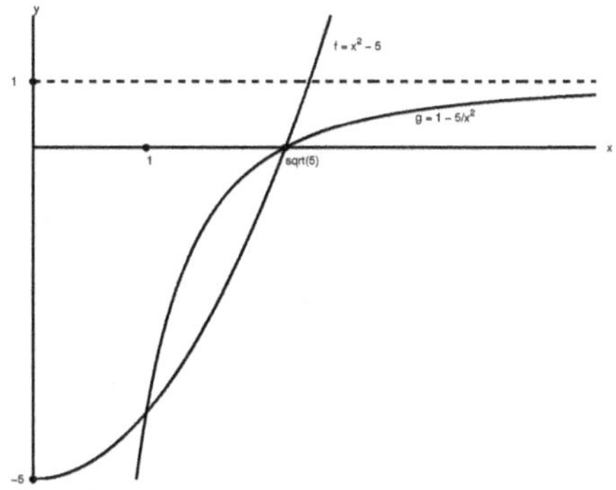

For x_0 near $m\sqrt{R}$, the two formulas behave similarly with the second being computationally simpler—no division by x_n.

However, the first formula is better since the second may not converge for x_0 near 0 or x_0 large >> $n\sqrt{R}$.

3.2.29. $f(x) = \arctan x$, $f'(x) = 1/(1+x^2)$.

So we have $x - \dfrac{f}{f'} = x - (1+x^2)\arctan x$.

Therefore, we obtain $x_{n+1} = x_n - (1+x_n^2)\arctan x_n$. $x_0 = 2$, $x_1 = -3.54$, $x_2 = 13.95$, $x_3 = -279.34$.

Newton's method diverges.

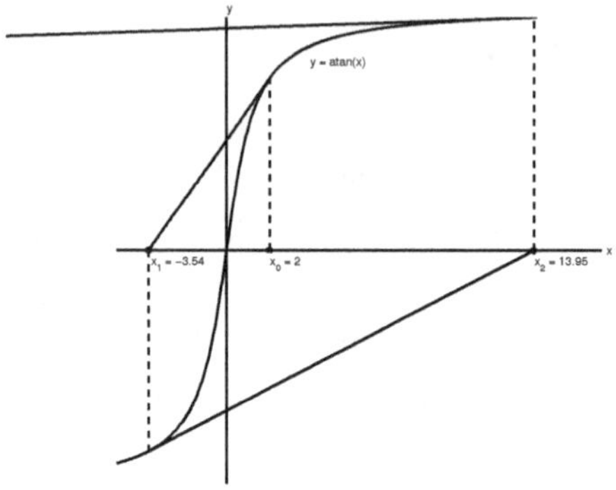

3.2.31. Given $F(x) = f(x)/f'(x)$. Now

$$F'(x) = \frac{f'(x)f'(x) - f(x)f''(x)}{[f'(x)]^2} = 1 - \frac{f(x)f''(x)}{[f'(x)]^2}$$

Hence, Newton's method for F is

$$x_{n+1} = x_n - \frac{F(x_n)}{F'(x_n)} = x_n - \frac{f(x_n)f'(x_n)}{[f'(x_n)]^2 - f(x_n)f''(x_n)}$$

By the results in the text, Newton's method for F will converge quadratically if F'' is continuous and $F'(r) \neq 0$.

The first condition is true because we assume f''' continuous and $f'(r) \neq 0$ so $f'(x) \neq 0$ for x near r.

The second condition is true because if $F'(r) = 0$ then $[f'(r)]^2 = f(r)f''(r) = 0$ contrary to hypothesis.

3.2.32. Retain three terms in the Taylor series:

$$0 = f(x+h) \approx f(x) + hf'(x) + \frac{h^2}{2}f''(x)$$

We solve the quadratic $\left(\frac{f''}{2}\right)h^2 + (f')h + (f) = 0$ for h and obtain

$$h = \left[-f' + \sqrt{(f')^2 - 2ff''}\right] \Big/ (f'')$$

Hence, we have

$$x_{n+1} = x_n + h = x_n - \frac{f'(x_n)}{f''(x_n)} + \frac{\sqrt{[f'(x_n)]^2 - 2f(x_n)f''(x_n)}}{f''(x_n)}$$

Let $e_{n+1} = x_{n+1} - r$ where r is a root of f. Then

$$e_{n+1} = x_n + h - r = e_n + h$$

Now

$$0 = f(r) = f(x_n - e_n) = f(x_n) - e_n f'(x_n) + \frac{e_n^2}{2}f''(x_n) - \frac{e_n^3}{6}f'''(\xi_n)$$

for some ξ_n. Moreover, we have

$$\begin{aligned}
0 &= f - (e_{n+1} - h)f' + \frac{1}{2}(e_{n+1} - h)^2 f'' - \frac{e_n^3}{6}f''' \\
&= [f + hf' + \frac{1}{2}h^2 f''] - e_{n+1}[f' + hf''] + \frac{1}{2}e_{n+1}^2 f'' - \frac{1}{6}e_n^3 f''' \\
&= -e_{n+1}[f' + hf''] + \frac{1}{2}e_{n+1}^2 f'' - \frac{1}{6}e_n^3 f'''
\end{aligned}$$

by the Taylor series above.

We ignore the e_{n+1}^2 term as it is very small and obtain

$e_{n+1} = -\left\{\frac{1}{6}f'''(\xi_n)/[f'(x_n) + hf''(x_n)]\right\} e_n^3.$

Thus, we have cubic convergence.

3.2.35. Let r be a root of $f(x)$ of multiplicity m, i.e., $f(r) = f'(r) = \cdots = f^{(m-1)}(r) = 0$ and $f^{(m)}(r) \neq 0$.

By Taylor series, we have

$$\begin{aligned} f(r+h) &= f(r) + hf'(r) + \cdots + \frac{h^{m-1}}{(m-1)!} f^{(m-1)}(r) + \frac{h^m}{m!} f^{(m)}(r) + \cdots \\ &= \frac{h^m}{m!} f^{(m)}(r) + \cdots \end{aligned}$$

So we have $f(x_n) = f(r + e_n) = \dfrac{e_n^m f^{(m)}(r)}{m!} + \dfrac{e_n^{m+1} f^{(m+1)}(\xi_n)}{(m+1)!}$.

Similarly, we obtain $f'(x_n) = f'(r + e_n) = \dfrac{e_n^{m-1} f^{(m)}(r)}{(m-1)!} + \dfrac{e_n^m f^{(m+1)}(\eta_n)}{m!}$.

As in the textbook, we have

$$\begin{aligned} e_{n+1} = x_{n+1} - r = x_n - r - \frac{mf(x_n)}{f'(x_n)} &= e_n - \frac{mf(x_n)}{f'(x_n)} \\ &= e_n - \frac{\dfrac{e_n^m f^{(m)}(r)}{(m-1)!} + \dfrac{e_n^{m+1} f^{(m+1)}(\xi_n)}{(m+1)(m-1)!}}{\dfrac{e_n^{m-1} f^{(m)}(r)}{(m-1)!} + \dfrac{e_n^m f^{(m+1)}(\eta_n)}{m!}} \\ &= \frac{\dfrac{e_n^m f^{(m)}(r)}{(m-1)!} + \dfrac{e_n^{m+1} f^{(m+1)}(\eta)}{m!} - \dfrac{e_n^m f^{(m)}(r)}{(m-1)!} - \dfrac{e_n^{m+1} f^{(m+1)}(\xi_n)}{(m+1)(m-1)!}}{e_n^{m-1}\left[\dfrac{f^{(m)}(r)}{(m-1)!} + \dfrac{e_n f^{(m+1)}(\eta_n)}{m!}\right]} \\ &= e_n^2 \left[\frac{\dfrac{f^{(m+1)}(\eta_n)}{m!} - \dfrac{f^{(m+1)}(\xi_n)}{(m+1)(m-1)!}}{\dfrac{f^{(m)}(r)}{(m-1)!} + \dfrac{e_n f^{(m+1)}(\eta_n)}{m!}}\right] \end{aligned}$$

Hence, we obtain quadratic convergence.

3.2.36. We have $x_{n+1} = x_n - f(x_n)/g(x_n)$, where $g(x) = [f(x + f(x)) - f(x)]/f(x)$.
Clearly, there are 2 function evaluations per step.
Consider Taylor series

$$f(x+h) = f(x) + hf'(x) + \frac{h^2}{2}f''(x) + \cdots f(x + f(x))$$
$$= f(x) + f(x)f'(x) + \frac{[f(x)]^2 f''(x)}{2} + \ldots$$

Now
$$\frac{f(x + (f(x)) - f(x)}{f(x)} = f'(x) + \frac{f(x)f''(x)}{2} + \ldots$$

Hence, we obtain $\frac{f(x + f(x)) - f(x)}{f(x)} \approx f'(x)$ if $f(x)$ is small.
Consequently, we obtain $g(x) \approx f'(x)$.
Steffensen's iteration is Newton's method with an approximation for $f'(x)$.
Does not require knowledge of $f'(x)$; however, same number of function evaluations.
Consider $e_{n+1} = x_{n+1} - r = e_n - \frac{f}{g} = \frac{e_n g - f}{g}$ $0 = f(r) = f(x_n - e_n) =$
$f(x_n) - e_n f'(x_n) + \frac{1}{2}e_n^2 f''(\xi_n)$.
So we have $f - e_n g = -\frac{1}{2}e_n^2 f''$
Therefore we obtain $e_{n+1} = \frac{1}{2}e_n^2 \frac{f''}{g}$ and quadratic convergence.

3.2.37. $g(x) = x - \omega f/f'$, $g' = 1 - \omega[(f')^2 - ff']/(f')^2 = [(f')^2(1-\omega) + ff'\omega]/(f')^2$.
Now $g'(r) = (1 - \omega)$ since $f(r) = 0$, $f'(r) \neq 0$.
Hence, we obtain $|g'(r)| < 1$ if $0 < \omega < 2$.

3.2.41. Newton $x_1, x_2, x_3, x_4, \cdots$.
New method $x_2, x_4, \cdots$. $e_{n+1} \approx ce_n^2$ for Newton method.
Thus, we have $e_{n+2} \approx ce_{n+1}^2 \approx c(ce_n^2)^2 = c^3 e_n^4$.
It is 4th order.

3.2 Computer Exercises

3.2.C4. 0.32796 77853 31818 36223 77546

3.2.C5. 2.09455 14815 42326 59148 23865 40579

3.2.C8. 1.83928 67552

3.2.C9. 0.47033 169

3.2.C10. **a.** 1.89549 42670 340
b. 1.99266 68631 307
c. 2.58280 14730 552
d. 0.51097 34293 8857

3.2.C14. 3.13108, 3.15145 (two nearby roots)

3.3 Secant Method

3.3.1. $f(x) = x^4 - 4^3, x_{n+1} = x_n - f(x_n)\Big(\dfrac{x_n - x_{n-1}}{f(x_n) - f(x_{n-1})}\Big)$. $x_0 = 3$; $x_1 = 2$;

$$x_2 = 2 - (-48)\Big(\frac{2-3}{-48-17}\Big) = 2 + \frac{48}{65} = 2.7385$$

3.3.3. $f(x) = x^5 + x^3 + 3$. $x_{n-2} = 0,\ f(x_{n-2}) = 2;\ x_{n-1} = 1,\ f(x_{n-1}) = 5$:

$$x_n = x_{n-1} - f(x_{n-1})\left(\frac{x_{n-1} - x_{n-2}}{f(x_{n-1}) - f(x_{n-2})}\right) = 1 - 5\Big(\frac{1-0}{5-2}\Big) = -\frac{2}{3}$$

3.3.4. $x_{n+1} = x_n + (2 - e^{x_n})\left(\dfrac{x_n - x_{n-1}}{e^{x_n} - e^{x_{n-1}}}\right)$; $f(x) = e^x - 2 = 0,\ x = \ln 2$

3.3.9. $f(x) = x^2 - R$: $x_{n+1} = x_n - (x_n^2 - R)\left[\dfrac{x_n - x_{n-1}}{x_n^2 - x_{n-1}^2}\right] = x_n - \dfrac{x_n^2 - R}{x_n + x_{n-1}}$

3.3.12. By Taylor series, we have

$$0 = f(r) = f(x_n - e_n) = f(x_n) - e_n f'(\xi_n)$$

for some ξ_n. Thus, we have

$$e_{n+1} = e_n - \Big(\frac{x_n - x_0}{f(x_n) - f(x_0)}\Big)f(x_n) = e_n\Big[1 - \Big(\frac{x_n - x_0}{f(x_n) - f(x_0)}\Big)f'(\xi_n)\Big]$$

Hence, linear convergence can be assured if we suppose that $|1 - f'(u)/f'(v)| \le C < 1$ for all real u and v.

3.3.13. **a.** The sequence $1, \frac{1}{4}, \frac{1}{9}, \frac{1}{16}, \frac{1}{25}, \ldots, \frac{1}{n^2}$ has linear convergence since it converges to 0 and

$$e_n = \frac{1}{n^2} < \frac{1}{(n-1)^2} = e_{n-1}$$

d. The sequence $\frac{1}{4}, \frac{1}{16}, \frac{1}{256}, \ldots, \frac{1}{2^{2^n}}$ has quadratic convergence to 0 since

$$e_n = \frac{1}{2^{2^n}} = \Big(\frac{1}{2^{2^{n-1}}}\Big)^2 = e_{n-1}^2$$

3.3.15. Take $x_0 = 0$.

Since $|f(x) - f(x_0)| = |f'(z)||x| \le c|x|$ by the Mean-Value Theorem, the graph of f is in the shaded part.

So the line $y = x$ intersects the graph of f at some point, say $\xi = f(\xi)$.

Then $|\xi - x_{n+1}| = |f(\xi) - f(x_n)| = |f'(z)|\ |\xi - x_n| \le c|\xi - x_n|$.

Thus, the points get closer to ξ.

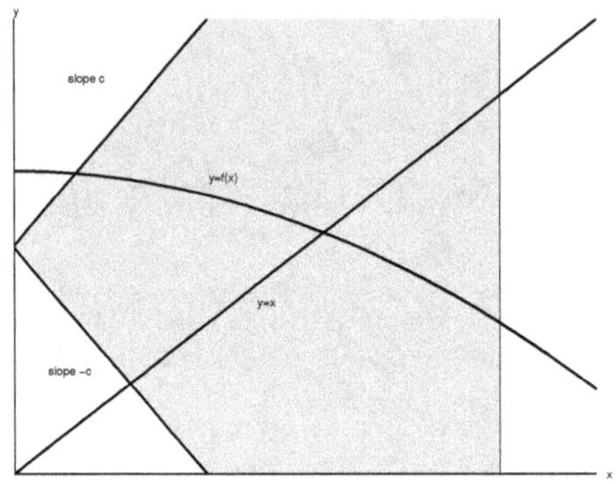

3.3.16. $f(x) = \frac{1}{2}x + \frac{1}{x}$: $x_0 = 1$, $x_1 = \frac{1}{2} + 1 = \frac{3}{2} = 1.5$, $x_2 = \frac{3}{4} + \frac{2}{3} = \frac{17}{12} = 1.467$, $x_3 = \frac{17}{24} + \frac{12}{17} = \frac{577}{408} = 1.4142$, $x_4 = \frac{577}{816} + \frac{408}{577} = 1.4142$.

Limit $\sqrt{2}$ same as Newton's method for $x_{n+1} = \frac{1}{2}x_n + \frac{1}{x_n}$.

3.3.17. $f = 5 + \frac{1}{2}\sin x$, $f' = \frac{1}{2}\cos x$, $|f'| \le \frac{1}{2} < 1$.

Thus, the method of functional iteration will produce a fixed point. $x_0 = \frac{3\pi}{2} = 4.7$, $x_1 = 5 + \frac{1}{2}\sin 4.7 = 4.50004$, $x_2 = 4.51123$, $x_3 = 4.51008$, $x_4 = 4.51012$, $x_5 = 4.510186$, $x_6 = 4.510187$, $x_7 = 4.510187$.

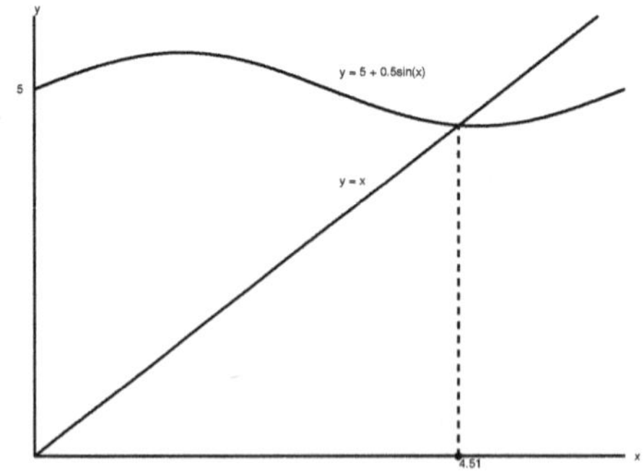

3.3 Computer Exercises

3.3.C1. -0.45896, 3.73308

3.3.C6. **a.** 1.53209

b. 1.23618

3.3.C7. 1.36880 81078 21373

3.3.C9. 20.80485 4

Chapter 4

Interpolation and Numerical Differentiation

4.1 Polynomial Interpolation

4.1.1. Since $p_{n-1}(x) = \sum_{i=0}^{n-1} f(x_i)\ell_i(x)$, we have

$$\begin{aligned} p_3(x) &= 7\frac{(x-2)(x-3)(x-4)}{(0-2)(0-3)(0-4)} \\ &+ 11\frac{(x-0)(x-3)(x-4)}{(2-0)(2-3)(2-4)} \\ &+ 28\frac{(x-0)(x-2)(x-4)}{(3-0)(3-2)(3-4)} \\ &+ 63\frac{(x-0)(x-2)(x-3)}{(4-0)(4-2)(4-3)} \end{aligned}$$

Hence, we obtain

$$\begin{aligned} p_3(x) &= -\frac{7}{24}(x-2)(x-3)(x-4) \\ &+ \frac{11}{4}(x)(x-3)(x-4) \\ &- \frac{28}{3}(x)(x-2)(x-4) \\ &+ \frac{63}{8}(x)(x-2)(x-3) \\ &= 7 - 2x + x^3 \end{aligned}$$

4.1.3. We consider $\ell_2(x)$ only since $\ell_0(x)$ and $\ell_1(x)$ are similarly done.

$$\begin{aligned}\ell_2(x) &= \frac{(x-x_0)(x-x_1)(x-x_3)}{(x_2-x_0)(x_2-x_1)(x_2-x_3)} \\ &= \frac{(x+1)(x-1)(x-4)}{(3+1)(3-1)(3-4)} \\ &= \frac{(x^2-1)(x-4)}{4\cdot 2\cdot(-1)} = -\frac{1}{8}(x-4)(x^2-1)\end{aligned}$$

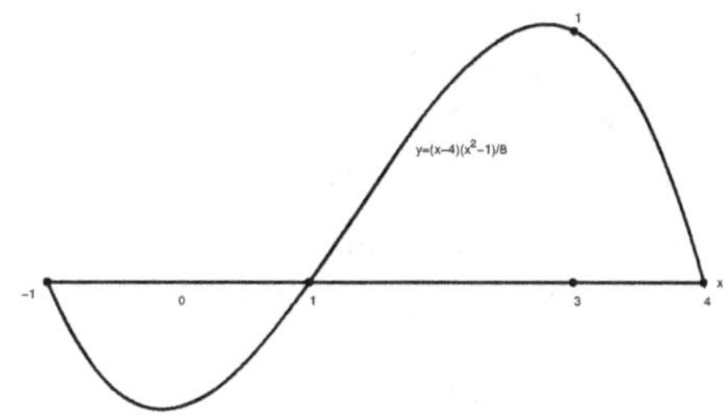

4.1.7. a.

−1	2			
		−3		
1	−4		2	
		5		−11/24
3	6		−3/4	
		2		
5	10			

$$\begin{aligned}p_3(x) &= 2-3(x+1)+2(x+1)(x-1)-\frac{11}{24}(x+1)(x-1)(x-3) \\ &= 2+(x+1)(-3+(x-1)(2+(x-3)(-11/24)))\end{aligned}$$

4.1.8.

1	−1				
		2/3			
2	−1/3		1/8		
		41/48		3/4	
2.5	3/32		39/24		11/6
		119/48		75/12	
3	4/3		113/8		
		71/3			
4	25				

$$\begin{aligned}p_4(x) &= -1+\frac{2}{3}(x-1)+\frac{1}{8}(x-1)(x-2)+\frac{3}{4}(x-1)(x-2)(x-2.5) \\ &\quad +\frac{11}{6}(x-1)(x-2)(x-2.5)(x-3) \\ &= -1+(x-1)(\frac{2}{3}+(x-2)(\frac{1}{8}+(x-2.5)(\frac{3}{4}+(x-3)\frac{11}{6})))\end{aligned}$$

4.1.9. **a.**

0	1				
		8			
1	9		3		
		14		$\boxed{1}$	
2	23		$\boxed{7}$		0
		$\boxed{35}$		1	
4	$\boxed{93}$		12		
		83			
6	259				

b. $p(x) \quad = 93 + 35(x-4) + 7(x-4)(x-2) + (x-4)(x-2)(x-1)$

$f(4.2) \approx p(4.2) = 93 + 35(0.2) + 7(0.2)(2.2) + 7(0.2)(2.2)(3.2) = 104.488$

We obtain the same polynomial using the either top diagonal downward or the second diagonal upward!

Notice that $93 + 35(x-4) + 7(x-4)(x-2) + (x-4)(x-2)(x-1) = 1 + 7x + x^3$ and $1 + 8(x) + 3(x)(x-1) + 1(x)(x-1)(x-2) = 1 + 7x + x^3$.

4.1.12. Clearly, $q(x) = p(x) + c(x+2)(x+1)(x)(x-1)(x-2)$.

When $x = 3$, $30 = 61 + c(5)(4)(3)(2)(1)$ and $c = -31/120$.

Thus, we have

$$q(x) = x^4 - x^3 + x^2 - x + 1 - \frac{31}{120}\,(x+2)(x+1)(x)(x-1)(x-2)$$

4.1.13. **a.**

0	−1				
		0			
1	−1		0		
		0		1	
2	−1		−1		0
		2		1	
−1	−7		1		
		3			
3	5				

$p_4(x) = -1 + (x)(x-1)(x-2) = x^3 - 3x^2 + 2x - 1$

4.1.14.

1	−1.5		
		1	
2	−0.5		0
		1	
2.5	0		0
		1	
3	0.5		0
		1	
4	1.5		

$p_4(x) = -1.5 + 1(x-1) = x - 2.5$

4.1.16. Let $p(x) = ax^2 + bx + c$. Then $p'(x) = 2ax + b$. Now by conditions $0 = c$, $1 = a + b$, $2 = 2a\alpha + b$. Hence, we obtain $a = 1/(2\alpha - 1)$ and $b = (2\alpha - 2)/(2\alpha - 1)$. So $p(x) = x(x + 2\alpha - 2)/(2\alpha - 1)$ with $\alpha \neq \alpha_0 = 1/2$.

4.1.18.

0	2				
		−1			
1	1		1		
		2		−2	
3	5		−3		−1
		−1		−7	
2	6		−31		
		−63			
5	−183				

$$p(x) = 2 - 1(x) + 1(x)(x-1) - 2(x)(x-1)(x-3) - 1(x)(x-1)(x-3)(x-2)$$

$$p(x) = 2 + x(-1 + (x-1)(1 + (x-3)(-2 + (x-2)(-1))))$$

4.1.19.

−2	−1				
		2			
−1	1		−1		
		0		1	
0	1		2		0
		4		1	
1	5		5		
		14			
2	19				

$$p_4(x) = -1 + 2(x+2) - (x+2)(x+1) + (x+2)(x+1)x$$

$$p_2(x) = 1 + 2(x+1)x$$

Remark: $p_4 = f$ by uniqueness of interpolation polynomial.

4.1.22.

1.73	0				
		0			
1.82	0		0		
		0		0	
5.22	0		0		0.76
		0		0.67	
8.26	0		0.529		
		−1.38			
2.61	7.8				

$$p(x) = 0.76(x - 1.73)(x - 1.82)(x - 5.22)(x - 8.26)$$

4.1.25.

2	1.5713			
		0.0006		
3	1.5719		0.00011667	
		0.00095		0.0
5	1.5738		0.00011667	
		0.0013		
6	1.5751			

$p_3(x) = 1.5713 + .0006(x-2) + .00011667(x-2)(x-3)$.

Clearly, $p_2(x) = p_3(x)$ and $p_2(4) = 1.5727$. No advantage!

4.1.27. Let $p(x) = ax^3 + bx^2 + cx + d$. Thus, we have $p'(x) = 3ax^2 + 2bx + c$. For the given conditions, we have

$$\begin{cases} 1 = d \\ 0 = a + b + c + d \\ 0 = c \\ -1 = 3a - 2b + c \end{cases}$$

Clearly, $a = -\frac{3}{5}$ and $b = -\frac{2}{5}$. Hence, we obtain $p(x) = -\frac{3}{5}x^3 - \frac{2}{5}x^2 + 1$.

4.1.28.

1.0	0.0					
		0.35218				
1.5	0.17609		−0.10230			
		0.24988		0.0265533		
2.0	0.30103		−0.0491933		−0.0064081	
		0.17609		0.0105333		0.001412
3.0	0.47713		−0.0281266		−0.002172	
		0.13390		0.0051033		
3.5	0.54407		−0.01792			
		0.11598				
4.0	0.60206					

$$p_3(x) = .35218(x-1) - .1023(x-1)(x-1.5) + .0265533(x-1)(x+1.5)(x-2)$$

$$q_3(x) = .17609 + .24988(x-1.5) - .0491933(x-1.5)(x-2) + .0105333(x-1.5)(x-2)(x-3)$$

$$p_3(1.2) = .077848, \ \log 1.2 = .07918; \quad q_3(2.4) = .38099, \ \log 2.4 = .3802$$

4.1.39. Forward-difference: $h = 0.01$, $x_0 = 2.35$, $s = 0.2$; $s = 3.7$

$$p_2(x_0 + sh) = f_0 + s\Delta f_0 + \tfrac{1}{2}s(s-1)\Delta^2 f_0$$

$$\ln 2.352 \approx 0.85442 + (0.2)(0.00424) + \tfrac{1}{2}(0.2)(-0.8)(-0.00001) \approx 0.85527$$

$$\ln 2.387 \approx 0.85442 + (3.7)(0.00424) + \tfrac{1}{2}(3.7)(2.7)(-0.00001) \approx 0.87006$$

$$\ln\ 2.352 = .855266, \text{ error } = 4 \times 10^{-6}$$

$$\ln\ 2.387 = .870034, \text{ error } = 26 \times 10^{-6}$$

4.1.40. Divisions: $\dfrac{1}{2^n}(n-1)$; additions/subtractions: $n(n-1)$

4.1.42. The Newton polynomial is:

$$p_n(x) = f[x_0] + f[x_0, x_1](x - x_0) + \cdots + f[x_0, x_1, \ldots, x_{n+1}](x - x_0)(x - x_1)\ldots(x - x_n)$$

Hence, we obtain $p[x_0, x, \ldots, x_{n+1}] = 0$.

4.1.45. False, only unique for polynomial p of degree $\leq n - 1$.

Consider $(0, y_0)$ and $(-1, y_0)$, then $p(x) = x^2 + x + y_0$ and $g(x) = 7x^2 + 7x + y_0$ have the desired property: $p(0) = y_0 = p(-1)$, $g(0) = y_0 = g(-1)$.

4.1 Computer Exercises

4.1.C1. $p(x) = 2 + 46(x-1) + 89(x-1)(x-2) + 6(x-1)(x-2)(x-3)$

$$+\ 4(x-1)(x-2)(x-3)(x-4)$$

4.2 Errors in Polynomial Interpolation

4.2.1.

-2	1					
		3				
-1	4		2			
		7		-1		
0	11		-1		0	
		5		-1		0
1	16		-4		0	
		-3		-1		
2	13		-7			
		-17				
3	-4					

The interpolating polynomial is of degree 3 since $f[x_0, x_1, x_2, x_3, x_4] = 0$.

4.2.6. The bound for linear interpolation is

$$\frac{1}{8}h^2 M$$

where $M = \max |f''(x)|$ over an interval of length h. Now $|f''(x)| = |\sin x| \leq 1$ and $h = 10^{-2}$ so the bound is $\leq \frac{1}{8}10^{-4} = 1.25 \times 10^{-5}$.

4.2.7. Linear interpolation:

$p_1(x) = f(x_0) + f[x_0, x_1](x - x_0) = f(x_0) + [(f(x_1) - f(x_0))/(x_1 - x_0)](x - x_0)$.

$$\begin{aligned}\sin(0.705) &\approx p_1(0.705)\\ &= \sin(0.7) + [(\sin(0.71) - \sin(.7))/(0.71 - 0.7)](0.705 - 0.7)\\ &= 0.64421\,76872 + [(0.0076160\,83784)/(0.01)](0.005)\\ &\approx 0.64421\,76872 + (0.0076160\,83784)(0.5)\\ &\approx 0.64802\,57291\\ \sin(0.705) &= 0.64803\ 38295\ldots \qquad (\text{Error} = 0.81 \times 10^{-5})\end{aligned}$$

$$\begin{aligned}\cos(0.702) &\approx p_1(0.72)\\ &= \cos(0.7) + [(\cos(0.72) - \cos(0.7))/(0.72 - 07)](0.702 - 0.7)\\ &\approx 0.76484\,21873 + [(-0.0064803\,4273)/(0.02)](0.002)\\ &\approx 0.76484\,21873 - (0.0064803\,11294)(0.2)\\ &\approx 0.76354\,61250\\ \cos(0.702) &= 0.76355\,22231\ldots \qquad (\text{Error} = 0.61 \times 10^{-5})\end{aligned}$$

Error bound is $\frac{1}{8}(0.01)^2 = 0.125 \times 10^{-4}$.

4.2.8. $\frac{1}{8}h^2M \le \frac{1}{8}h^2 < \frac{1}{2}10^{-5}$

$h < \sqrt{0.4} \times 10^{-2}$

$\frac{\pi}{n} < \sqrt{0.4} \times 10^{-2}$

$n > \frac{\pi}{\sqrt{0.4}} \times 10^2 = 496.7$

At least 497 table entries.

4.2.9. We are given $f(x) = e^{-x}$ and $n = 20$.

By Theorem 2, an upper bound on the error is $[1/(4(n+1)]M[(b-a)/n]^{n+1}$ where $|f^{n+1)}(x)| \le M$.

Since $M \le 1$, the maximum error is $[1/(4(21)](2/20]^{21} = (1/84)10^{-21} = 1.19 \times 10^{-23}$.

4.2.10. Since $f(x) = \ln x$, $f^{(n)}(x) = (-1)^{n-1}(n-1)!x^n$ and $|f^{(10)}(x)| \le 9!$ over $[1, 2]$.

Moreover, $|(x - x_0)\ldots(x - x_9)| \le (9!/4)h^{10}$ by example in text.

So $|f(x) - p(x)| = |\frac{f^{(10)}(x)}{10!}(x - x_0)\cdots(x - x_9)| \le \frac{9!}{40}h^{10}$.

But $h = \frac{1}{9}$ so the bound on the error is $9!(9^{-10})/40 \approx 2.6 \times 10^{-6}$.

4.2.13. Since $f(x) = \cos x$, we have $|f^{(n)}(x)| \leq 1$.

Then $|\text{error}| \leq |f(x) - p(x)| \leq |(x - x_0)(x - x_1) \dots (x - x_n)|/n!$.

But $|(x - x_0)(x - x_1) \dots (x - x_n)| \leq (n-1)! h^n/4$.

Thus, we have $|\text{error}| \leq h^n/(4n) = [4nn^n]^{-1}$ since $h = 1/n$.

For $n = 9$, accuracy is $1/(36 \times 9^9) \simeq 7.2 \times 10^{-11}$.

Require $[4nn^n]^{-1} < 10^{-7}$; $7 < \log 4 + \log n + n \log n$.

For $n = 7$, RHS $= 7.36$.

4.2.14. The largest bound occurs when x is in either outside subinterval.

Hence, we obtain

$$\prod_{i=0}^{n-1} |x - x_i| \leq \left(\frac{h}{2}\right)\left(\frac{3h}{2}\right)\left(\frac{5h}{2}\right)\cdots\left(\frac{(2n-1)h}{2}\right) = \left(\frac{h}{2}\right)^n (1 \cdot 3 \cdot 5 \cdots (2n-1))$$
$$= \frac{h^n (2n)!}{2^{2n} n!}$$

4.2.16. Put $q(x) = \sum_{i=0}^{n-1} p[x_0, \dots, x_i] \prod_{j=0}^{i-1} (x - x_j)$.

By the theory, q is a polynomial of degree $\leq n-1$ interpolating p at $x_0, \dots, x_n$.

So is p itself. By uniqueness, $p = q$.

4.2 Computer Exercises

4.2.C10. On $[-1, 1]$ we have interpolation error $f(x) - p(x) = \dfrac{f^{(n)}(\xi)}{n!}(x - x_0) \dots (x - x_n)$.

The term involving the nodes is $w(x) = (x - x_0) \dots (x - x_n)$.

The Chebyshev nodes are the nodes which minimize the "supreme norm" of w on $[-1, 1]$:

$$||w|| = \max_{-1 \leq x \leq 1} |(x - x_0)(x - x_1) \dots (x - x_n)|$$

So they make the error formula most favorable.

4.3 Estimating Derivatives and Richardson Extrapolation

4.3.1. $f(x+3h) = f(x) + (3h)f'(x) + \dfrac{(3h)^2}{2!}f''(x) + \cdots$

$f(x-h) = f(x) - hf'(x) + \dfrac{h^2}{2!}f''(x) + \cdots$

Therefore: $f'(x) = \dfrac{1}{4h}[f(x+3h) - f(x-h)] - hf''(x) + \cdots$

Error term $= -hf''(\xi)$ for $\xi \in (x-h, x+3h)$

4.3.2. $f(2h) = f(0) + 2hf'(0) + \dfrac{(2h)^2}{2!}f''(\xi)$ for $\in (0, 2h)$

$f(2h) - f(0) = 2hf'(0) + 2h^2 f''(\xi)$

$f'(0) = \dfrac{1}{2h}[f(2h) - f(0)] - hf''(\xi)$

Error term $= -hf''(\xi)$ for $\xi \in (0, 2h)$

4.3.4. The answer is that no such formula exists. To prove: Suppose

$$f'(x) = \frac{\alpha f(x+\lambda h) + \beta f(x+\mu h)}{h} + \mathcal{O}(h^3)$$

where $\alpha\lambda\beta\mu$ are given real numbers. Then

$$\begin{aligned} f'(0) &= \frac{\alpha}{h}f(\lambda h) + \tfrac{\beta}{h}f(\mu h) + \mathcal{O}(h^3) \\ f'(0) &= \frac{\alpha}{h}\left[f(0) + \lambda h f'(0) + \tfrac{1}{2}(\lambda h)^2 f''(0) + \tfrac{1}{6}(\lambda h)^3 f'''(0) + \mathcal{O}(h^4)\right] \\ &\quad + \frac{\beta}{h}\left[f(0) + \mu h f'(0) + \tfrac{1}{2}(\mu h)^2 f''(0) + \tfrac{1}{6}(\mu h)^3 f'''(0) + \mathcal{O}(h^4)\right] + \mathcal{O}(h^3) \end{aligned}$$

All terms of order h^{-1}, h^0, h^1, h^2 must disappear. Hence, we obtain

$$\begin{cases} h^{-1} : \alpha + \beta = 0 \\ h^0 \;\; : \alpha\lambda + \beta\mu = 1 \\ h^1 \;\; : \alpha\lambda^2 + \beta\mu^2 = 0 \\ h^2 \;\; : \alpha\lambda^3 + \beta\mu^3 = 0 \end{cases}$$

Eliminate β to get

$$\begin{cases} \alpha\lambda - \alpha\mu = 1 \\ \alpha\lambda^2 - \alpha\mu^2 = 0 \\ \alpha\lambda^3 - \alpha\mu^3 = 0 \end{cases}$$

Eliminate α to get

$$\begin{cases} \dfrac{1}{\lambda - \mu}(\lambda^2 - \mu^2) = 0 \\ \dfrac{1}{\lambda - \mu}(\lambda^3 - \mu^3) = 0 \end{cases}$$

The last equations simplify to $\lambda + \mu = \lambda^2 + \lambda\mu + \mu^2 = 0$. This forces $\mu = \lambda = 0$. So system is inconsistent.

4.3.6. The point ξ for the first Taylor series is such that $\xi \in (x, x+h)$ while the second is $\xi \in (x-h, x)$. Clearly, they are not the same.

4.3.8. $f(x+2h) = f + 2hf' + 2h^2 f'' + \frac{4}{3}h^3 f''' + \frac{2^4}{4!}h^4 f^{(4)} + \cdots$

$f(x-2h) = f - 2hf' + 2h^2 f'' - \frac{4}{3}h^3 f''' + \frac{2^4}{4!}h^4 f^{(4)} + \cdots$

Here $f = f(x), f' = f'(x), f'' = f''(x)$ and so on.

a. $f(x+2h) - f(x-2h) = 4hf' + \frac{8}{3}f''' + \cdots$

$$f'(x) = [f(x+2h) - f(x-2h)]/(4h) - \frac{2}{3}h^2 f''(x) + \cdots$$

The error term is $-\frac{2}{3}h^2 f''(\xi)$.

4.3.9. $D(h) \equiv f(x+h) - f(x-h) = 2hf' + h^3 f'''/3 + 2h^5 f^{(5)}/5! + \cdots$

$S(h) \equiv f(x+h) + f(x-h) = 2f + h^2 f'' + 2h^4 f^{(4)}/4! + 2h^6 f^{(6)}/6 + \cdots$

Here $f = f(x), f' = f'(x), f'' = f''(x)$ and so on.

a. $D(2h) = 4hf' + 2^3 h^3 f'''/3 + 2^6 h^5 f^{(5)}/5! + \cdots$

$2D(h) = 4hf' + 2h^3 f'''/3 + 2^2 h^5 f^{(5)}/5! + \cdots$

$D(2h) - 2D(h) = 2h^3 f''' + \frac{60}{120}h^5 f^{(5)} + \cdots$

$$f'''(x) = [D(2h) - 2D(h)]/(2h^3) - \frac{h^2}{4}f^{(5)}(x) + \cdots$$

The error term is $-\frac{h^2}{4}f^{(5)}(\xi)$.

b. $S(h) = 2f + h^2 f'' + h^4 f^{(4)}/12 + h^6 f^{(6)}/360 + \cdots$

$S(2h) = 2f + 4hf'' + 16h^4 f^{(4)}/12 + 64h^6 f^{(6)}/360 + \cdots$

$S(2h) - 4S(h) = -6f + h^4 f^{(4)} + h^6 f^{(6)}/6 + \cdots$

$S(2h) - 4S(h) + 6f = h^4 f^{(4)} + \frac{h^6}{6}f^{(6)} + \cdots$

$$f^{(4)}(x) = [S(2h) - 4S(h) + 6f]/h^4 - \frac{h^2}{6}f^{(6)}(x) + \cdots$$

The error term is $-\frac{h^2}{6}f^{(6)}(\xi)$.

Both $\mathcal{O}(h^2)$ but second has smaller error term.

4.3.11. $f(x_0) = f(x_1 - h) = f - hf' + \frac{h^2}{2} f'' - \frac{h^3}{3!} f''' + \frac{h^4}{4!} f^{(4)} - \cdots$

$f(x_2) = f(x_1 + \alpha h) = f + \alpha h f' + \frac{\alpha^2 h^2}{2} f'' + \frac{\alpha^3 h^3}{3!} f''' + \frac{\alpha^4 h^4}{4!} f^{(4)} + \cdots$

Thus, we have

$$f(x_2) - f(x_0) = (\alpha + 1) h f' + (\alpha^2 - 1)\ \frac{h^2}{2} f'' + (\alpha^3 + 1) \frac{h^3}{3!} f'''.$$

Since $x_2 - x_0 = (1 + \alpha) h$, we have

$$f'(x_1) = \frac{f(x_2) - f(x_0)}{x_2 - x_0} - (\alpha - 1)\frac{h}{2} f''(x_1) - (\alpha^2 - \alpha + 1)\ \frac{h^2}{6} f'''(x_1) + \cdots .$$

When x_1 is the midpoint, $\alpha = 1$ and the error term is $-\frac{h^2}{6} f'''(\xi) = \mathcal{O}(h^2)$.

For any $\alpha \neq 1$, the error term is $-(\alpha - 1)\frac{h}{2} f''(\xi) = \mathcal{O}(h)$.

4.3.12. $f(x + h) = f(x) + h f'(x) + \frac{h^2}{2} f''(x) + \frac{h^3}{6} f'''(x) + \frac{h^4}{24} f^{(4)}(x) + \cdots$

$f(x - h) = f(x) - h f'(x) + \frac{h^2}{2} f''(x) - \frac{h^3}{6} f'''(x) + \frac{h^4}{24} f^{(4)}(x) + \cdots$

So we have

$$f(x + h) - f(x - h) = 2h f'(x) + \frac{h^3}{3} f'''(x) + \frac{h^5}{60} f^{(5)}(x) + \cdots$$

$$\frac{f(x + h) - f(x - h)}{2h} = f'(x) + \frac{h^2}{6} f'''(\xi_1)$$

Similarly, we have

$$\frac{f(x + h) - 2f(x) + f(x - h)}{h^2} = f''(x) + \frac{h^2}{12} f^{(4)}(\xi_2)$$

$$f(x + h) \left[\frac{1}{2h} + \frac{1}{h^2}\right] - 2f(x) \left[\frac{1}{h^2}\right] + f(x - h) \left[-\frac{1}{2h} + \frac{1}{h^2}\right]$$

$$= f'(x) + f''(x) + h^2 \left[\frac{f'''(\xi_1)}{6} + \frac{f^{(4)}(\xi_2)}{12}\right]$$

Hence, we obtain $f'(x) + f''(x) \approx \left(\frac{h + 2}{2h^2}\right) f(x + h) - \left(\frac{2}{h^2}\right) f(x) + \left(\frac{2 - h}{2h^2}\right) f(x - h)$

with error term $-\frac{h^2}{12}[2f''(\xi_1) + f^{(4)}(\xi_2)] = \mathcal{O}(h^2)]$ for $\xi_1, \xi_2 \in (x - h, x + h)$.

4.3.13.

x_0	f_0		
		$a = \dfrac{f_1 - f_0}{x_1 - x_0}$	
x_1	f_1		$c = \dfrac{b - a}{x_2 - x_0}$
		$b = \dfrac{f_2 - f_1}{x_2 - x_1}$	
x_2	f_2		

Polynomial is $p(x) = f_0 + a(x - x_0) + c(x - x_0)(x - x_1)$.

Thus, we have

$$p'(x) = a + c[2x - (x_0 + x_1)]$$

and

$$p'\left(\frac{x_0 + x_1}{2}\right) = a + c[(x_0 + x_1) - (x_0 + x_1)] = a = \frac{f_1 - f_0}{x_1 - x_0}$$

4.3.16. $2\varphi\left(\frac{h}{2}\right) - \varphi(h) = L + \frac{1}{2}c_2h^2 + \frac{3}{4}c_3h^3 + \cdots$

$$L \approx 2\varphi\left(\frac{h}{2}\right) - \varphi(h)$$

4.3.20. Now

$$L - \varphi(h) \approx ce^{-1/h} \quad \Rightarrow c \approx [L - \varphi(h)]e^{1/h}$$

$$L - \varphi\Big(\frac{h}{2}\Big) \approx ce^{-2/h} \Rightarrow c \approx [L - \varphi\Big(\frac{h}{2}\Big)]e^{2/h}$$

$$L - \varphi\Big(\frac{h}{3}\Big) \approx ce^{-3/h} \Rightarrow c \approx [L - \varphi\Big(\frac{h}{3}\Big)]e^{3/h}$$

Equating the first two we have $\dfrac{L - \varphi(h)}{L - \varphi(h/2)} = e^{1/h}$.

From this and the second expression above, we have

$$c = \Big[L - \varphi\Big(\frac{h}{2}\Big)\Big]\Big[\frac{L - \varphi(h)}{L - \varphi(h/2)}\Big]^2 = \frac{[L - \varphi(h)]^2}{[L - \varphi(h/2)]}$$

Finally, from the third expression, we have

$$L \approx \varphi\Big(\frac{h}{3}\Big) + \frac{[L - \varphi(h)]^2}{[L - \varphi(h/2)]}\Big[\frac{L - \varphi(h/2)}{L - \varphi(h)}\Big]^3 = \varphi\Big(\frac{h}{3}\Big) + \frac{[L - \varphi(h/2)]^2}{[1 - \varphi(h)]}$$

Hence, we obtain

$$L[L - \varphi(h)] = \varphi\Big(\frac{h}{3}\Big)[L - \varphi(h)] + \Big[L - \varphi\Big(\frac{h}{3}\Big)\Big]^2$$

Solving for L, we have

$$L = \frac{[\varphi(h/2)]^2 - \varphi(h)\varphi(h/3)}{2\varphi(h/2) - \varphi(h) - \varphi(h/3)}$$

Alternative Approach: Notice that

$$(ce^{-2/h})^2 = (ce^{-1/h})(ce^{-3/h})$$

so

$$(L - \varphi(h/2))^2 = (L - \varphi(h))(L - \varphi(h/3))$$

or

$$L^2 - 2L\varphi(h/2) + [\varphi(h/2)]^2 = L^2 - \varphi(h)L - \varphi(h/3)L + \varphi(h)\varphi(h/3)$$

$$L[\varphi(h/3) - 2\varphi(h/2) + \varphi(h)] = -[\varphi(h/2)]^2 + \varphi(h)\varphi(h/3)$$

Hence, we obtain

$$L = \frac{[\varphi(h/2)]^2 - \varphi(h)\varphi(h/3)}{-\varphi(h/3) + 2\varphi(h/2) - \varphi(h)}$$

4.3 Computer Exercises

4.3.C3. 0.2021158503

Chapter 5

Numerical Integration

5.1 Trapezoid Rule

5.1.1. $\int_0^1 dx/(x^2+2) \approx \frac{1}{4}[f(0)+f(1)] + \frac{1}{2}f(\frac{1}{2}) = \frac{1}{4}[\frac{1}{2}+\frac{1}{3}] + \frac{1}{2}\left(\frac{4}{9}\right) = \frac{31}{72} \approx 0.4306$

5.1.3. Using $e^x = 1 + x + \frac{x^2}{2} + \frac{x^3}{3!} + \cdots$, we have $(e^x - 1)/x = 1 + \frac{x}{2} + \frac{x^2}{6} + \cdots$.

Thus, we have

$$\int_0^\alpha \left(\frac{e^x - 1}{x}\right) dx = \alpha + \frac{\alpha^2}{4} + \frac{\alpha^3}{18} + \cdots$$

$$\approx 10^{-4} + 0.25 \times 10^{-8} + 0.0567 \times 10^{-12} \approx 0.00010000250006$$

5.1.8.

$$T(f;P) = \frac{1}{2}\sum_{i=0}^{n-1}(x_{i+1} - x_i)[f(x_i) + f(x_{i+1})]$$

$$= \frac{1}{2}\left[\left(\frac{4}{3} - 1\right)\left(1 + \frac{3}{4}\right) + \left(2 - \frac{4}{3}\right)\left(\frac{1}{2} + \frac{3}{4}\right)\right]$$

$$= \frac{17}{24} \approx 0.70833$$

5.1.9. $T(f;P) = \frac{h}{2}[f(0) + f(1)] + hf(\frac{1}{2})$ where $f(x) = (x^2+1)^{-1}$ and $h = \frac{1}{2}$

$$= \frac{1}{4}[1 + \frac{1}{2}] + \frac{1}{2}\left(\frac{4}{5}\right) = \frac{31}{40} = 0.775$$

$$\int_0^1 \frac{dx}{x^2+1} = \arctan x\Big|_0^1 = \frac{\pi}{4} \approx 0.7854$$

$$\text{Error} = -\frac{1}{12}(b-a)h^2 f''(\xi) = -\frac{1}{48}f''(\xi)$$

$$f(x) = (x^2+1)^{-1},\ f'(x) = -(x^2+1)^{-2}(2x),$$

$$f''(x) = -2(x^2+1)^{-2} + 2(x^2+1)^{-3}(2x)^2$$

$$= 2(3x^2-1)/(x^2+1)^3$$

$$f'''(x) = [(x^2+1)^3(12x) - 2(3x^2-1)(3)(x^2+1)^2(2x)]/(x^2+1)^6$$

$$= (x^2+1)^2(12x)[x^2+1-3x^2+1]/(x^2+1)^6$$

$$= -24x(x^2-1)(x^2+1)^2/(x^2+1)^6$$

Set $f'''(x) = 0$. The maximum value occurs at $x = 1$ with $f''(1) = 1/2$.

An upper bound of the error term is $-[(b-a)/12]h^2 f''(1) = 1/96$.

Note that $(0.7854 - 0.775) = 0.0104 \le \frac{1}{96} = 0.01042$.

5.1.13. For $\int_a^b f(x)\,dx$ with uniform spacing $h = (b-a)/n$.

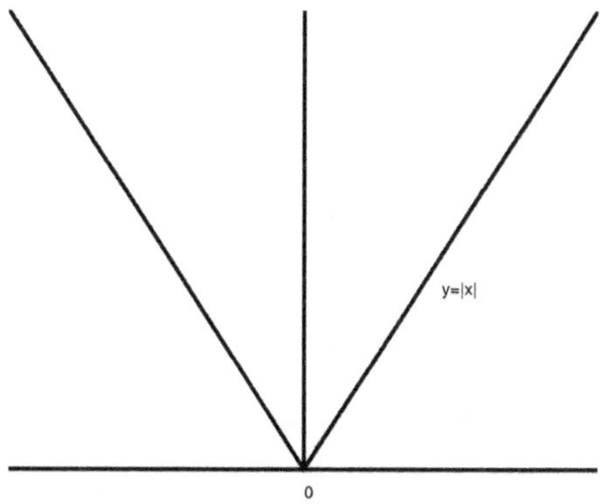

So the number of points involved is $n = (b-a)/h$. Clearly, we have

h	2	1	1/2	1/4
T	2	1	1	1

Since $f(x) = |x|$, we have

$$\int_{-1}^1 |x|\,dx = \int_{-1}^0 (-x)\,dx + \int_0^1 x\,dx = 2\int_0^1 x\,dx = 1$$

5.1.14. Error $= \dfrac{b-a}{12} h^2 f''(\xi)$. Now $\max\limits_{a \le x \le b} |f''(x)| = \max\limits_{0 \le x \le \pi} |-\sin x| \le 1$, and $h = (b-a)/n$. Hence, we obtain $|\text{Error}| \le (\pi/12)(\pi/n)^2 \le 10^{-12}$. So $n \ge 10^6 \pi \sqrt{\pi/12} = 1,607,438$. Too small since $\sin x$ is concave downward as shown.

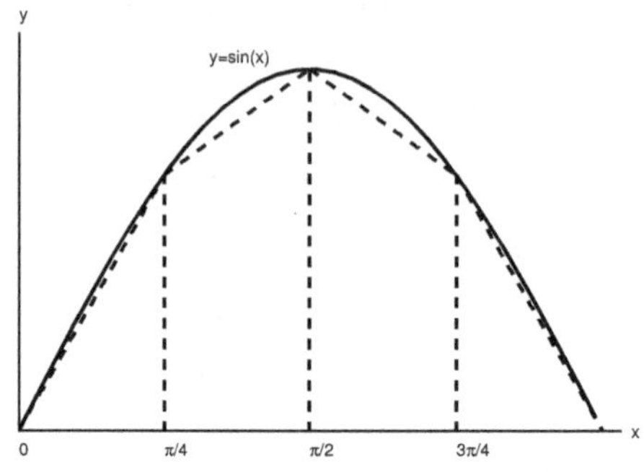

5.1.15. By (1) in the text, we have

$$T = h \sum_{i=1}^{n-1} x_i^2 + \frac{h}{2}[x_0^2 + x_n^2] = \frac{1}{n^3}(1^2 + 2^2 + 3^2 + \cdots + (n-1)^2) + \frac{1}{2n}$$

since $x_0 = 0$, $x_n = 1$, $h = n^{-1}$, $x_i = i/n$

$$T = \frac{1}{n^3}[\frac{1}{6}(n-1)(2n-1)(n)] + \frac{1}{2n}$$

Now

$$\lim_{n \to \infty} T = \lim_{n \to \infty} \left(\frac{1}{3} + \frac{1}{6n^2} \right) = \frac{1}{3}$$

5.1.19. $|\text{Error}| = \dfrac{1}{12}(b-a)h^2 |f''(\xi)| = \dfrac{1}{12}(5-2)(0.01)^2 |\sin(\xi)|$

$\le \dfrac{1}{12}(3)(0.01)^2 = .000025$

5.1.20. $T(f;P) = \dfrac{1}{4}[1+4] + \dfrac{1}{2}[2^{\frac{1}{2}} + 2^1 + 2^{\frac{3}{2}}] = \dfrac{9}{4} + \dfrac{3}{2}\sqrt{2} \approx 4.37132$

5.1.21. $T(f;P) = \dfrac{1}{4}\left[\dfrac{1}{2} + \dfrac{1}{3}\right] + \dfrac{1}{2}\left(\dfrac{4}{9}\right) = \dfrac{5}{24} + \dfrac{2}{9} = \dfrac{21}{72} \approx 0.43056$

5.1.22. $f(x) = \cos x^3$ Error term $= -\frac{1}{12}(b-a)h^2 f''(\xi)$

$f'(x) = -3x^2 \sin x^3,\ f''(x) = -6x \sin x^3 - 9x^4 \cos x^3$

$|f''(x)| = |-6x \sin x^3 - 9x^4 \cos x^3| \le 6x|\sin x^3| + 9x^4|\cos x^3| \le 6x + 9x^4$

$\max_{0 \le x \le 4} |f''(x)| \le 6(4) + 9(4^4) = 2328$

$|\text{ Error term }| \le \frac{1}{12}(4-0)\left(\frac{4}{200}\right)^2 2328 = \frac{2324}{7500} = .3104$

5.1.23. $T(f;P) = \frac{1}{8}[10+5] + \frac{1}{4}[8+7+6] = \frac{15}{8} + \frac{21}{4} = \frac{57}{8} = 7.125$

The upper and lower sums can not be used given only a table of function values since the extrema of the function would not be known.

5.1.24. $\int_a^{a+h} f(x)\,dx = hf + \frac{1}{2}h^2 f' + \frac{1}{6}h^3 f'' + \cdots$ where $f = f(a), f' = f'(a), f'' = f''(a), \ldots$.

a. By Taylor series, we have

$$hf(a+h) = h[f + hf' + \frac{1}{2}h^2 f'' + \frac{1}{6}h^3 f''' + \cdots] = hf + h^2 f' + \frac{1}{2}h^3 f'' + \frac{1}{6}h^4 f''' + \cdots$$

Hence, we obtain $\int_a^{a+h} f(x)\,dx - hf(a+h) = -\frac{1}{2}h^2 f'(\xi)$ for some $\xi \in (a, a+h)$.

Now $\int_a^b f(x)\,dx = \sum_{i=1}^n \int_{x_{i-1}}^{x_i} f(x)\,dx = h\sum_{i=1}^n f(x_i) - \frac{1}{2}h^2 \sum_{i=1}^n f'(\xi_i)$.

Error term is $-\frac{(nh)h}{2}\left[\frac{1}{n}\sum_{i=1}^n f'(\xi_i)\right] = -\frac{(b-a)h}{2} f'(\xi)$ for some $\xi \in (a,b)$.

b. Clearly, we have

$$\int_a^{a+h} f(x)\,dx - hf(a+h) = -\frac{1}{2}h^2 f'(a) - \frac{1}{6}h^3 f''(\xi)$$

for some $\xi \in (a, a+h)$. Now the error term for the general rule is

$$-\frac{(nh)h^2}{6}\left[\frac{1}{n}\sum_{i=1}^n f''(\xi_i)\right] = -\frac{(b-a)h^2}{6} f''(\xi)$$

for some $\xi \in (a,b)$.

5.1.25. **a.** By Taylor series, we have $hf(a+\frac{1}{2}h) = h[f + \frac{1}{2}hf' + \frac{1}{8}h^2 f'' + \cdots]$
$= hf + \frac{1}{2}h^2 f' + \frac{1}{8}h^3 f''' + \cdots$ where $f = f(a), f' = f'(a), f'' = f''(a), \ldots$.

Moreover, we have $\displaystyle\int_a^{a+h} f(x)\,dx = hf + \frac{1}{2}h^2 f' + \frac{1}{6}h^3 f'' + \cdots$.

Hence, we obtain $\displaystyle\int_a^{a+h} f(x)\,dx - hf(a + \frac{1}{2}h) = \tfrac{1}{24}h^3 f'' + \cdots$.

Error term is $\dfrac{h^3}{24}f''(\xi)$ for some $\xi \in (a, a+h)$.

b. For n-unequal subintervals, we have

$$\int_a^b f(x)\,dx = \sum_{i=1}^n \int_{x_{i-1}}^{x_i} f(x)\,dx \approx \sum_{i=1}^n h_i f(x_{i-1} + \frac{1}{2}h_i) + \sum_{i=1}^n \Big(\frac{h_i^3}{24} f''(\xi_i)\Big)$$

The error term is $\displaystyle\frac{1}{24}\sum_{i=1}^n h_i^3 f''(\xi_i)$.

c. For n-equal subintervals, $h = h_i$ for all i. Thus, the error term is

$$\frac{nh^3}{24}\left[\frac{1}{n}\sum_{i=1}^n f''(\xi_i)\right] = \frac{(b-a)}{24}h^2 f''(\xi)$$

for some $\xi \in (a, b)$ by the Intermediate Value Theorem for Continuous Functions.
Notice that $x_{i-1} + \frac{1}{2}h_i = a + ih + \frac{1}{2}h = a + (i + \frac{1}{2})h$.

5.1.29. Let $f(x) = x^n$ $(n > 3)$ on $[0, 1]$, with partition $\{0, 1\}$. $T = \frac{1}{2}$, $L = 0$, true value $= \int_0^1 x^n\,dx = \dfrac{1}{(n+1)}$. As an estimate of $\dfrac{1}{n+1}$, 0 is better than $\frac{1}{2}$ when $|0 - \dfrac{1}{n+1}| < |\frac{1}{2} - \dfrac{1}{n+1}|$ or $\dfrac{1}{n+1} < \frac{1}{4}$. Thus, we have $n > 3$.

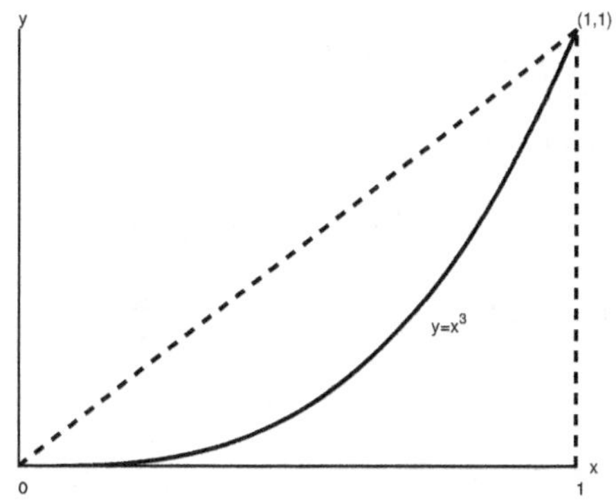

5.1.30. Since f is convex, it lies below the chords, as shown. Hence, we obtain

$$\int_a^b f(x)\,dx \leq T$$

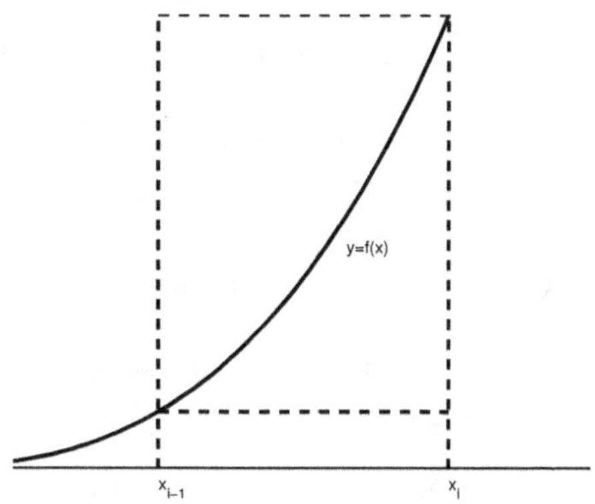

5.1.31. Error term $= \dfrac{b-a}{12} h^2 f''(\xi)$ for some $\xi \in (a, b)$. Now $f(x) = e^{-x^2}$, $f'(x) = -2xe^{-x^2}$, $f''(x) = -2e^{-x^2} + 4x^2 e^{-x^2}$. Thus, we have

$$\max_{0\leq x\leq 2} |f''(x)| \leq 2 \max_{0\leq x\leq 2} e^{-x^2} |2x^2 - 1| = 2$$

and

$$|\text{Error}| \leq \frac{1}{6} h^2 (2) = \frac{1}{3} h^2 = \frac{4}{3n^2} < 10^{-6}$$

since $h = 2/n$. Hence, we obtain $n > (2/\sqrt{3})\ 10^3$ or $n \geq 1155$. Note true solution 0.8820139.

5.1.34. For uniform spacing, $x_i - x_{i-1} = h$ and $x_i = a + ih$. So

$$R - \sum_{i=1}^{n}(x_i - x_{i-1})f(x_i) = h[f(a+h) + \cdots + f(b)]$$

Consider $\int_a^{a+h} f(x)\,dx \approx hf(a+h)$ and we have

$$(A)\ \int_a^{a+h} f(x)\,dx = hf(a) + \frac{h^2}{2}f'(a) + \frac{h^3}{3!}f''(a) + \cdots + \frac{h^n}{n!}f^{(n-1)}(a) + \cdots$$

By Taylor series,

$$(B)\ hf(a+h) = hf(a) + h^2 f'(a) + \frac{h^3}{2!}f''(a) + \cdots + \frac{h^n}{(n-1)!}f^{(n-1)}(a) + \cdots$$

Subtracting (A) and (B), we have

$$\int_a^{a+h} f(x)\,dx - hf(a+h) = -\frac{h^2}{2}f'(a) - \frac{h^3}{3}f''(a) - \cdots - \frac{h^n}{n!}(n-1)f^{(n-1)}(a) + \cdots$$

Thus, we have

$$(C)\ \int_a^{a+h} f(x)\,dx = hf(a+h) - \frac{h^2}{2}f'(\xi)$$

for some ξ such that $a < \xi < a + h$. Now

$$\int_a^b f(x)\,dx = \sum_{i=1}^{n}\int_{x_{i-1}}^{x_i} f(x)\,dx = \sum_{i=1}^{n}[hf(x_i) - \frac{h^2}{2}f'(\xi_i)]$$

using (C) with $x_{i-1} = a$ and $x_i = a + h$. Here $x_{i-1} < \xi_i < x_i$. Thus, we have

$$\int_a^b f(x)\,dx = h\sum_{i=1}^{n} f(x_i) - \left(\frac{h^2 n}{2}\right)\left[\frac{1}{n}\sum_{i=1}^{n} f(\xi_i)\right]$$

so that the error term is $-\frac{1}{2}(b-a)hf'(\xi) = \mathcal{O}(h)$, for $a < \xi < b$, by Intermediate Value Theorem for Continuous Functions and nh

5.1.35.
$$\begin{aligned}
\int_a^b f(x)\,dx &= \sum_{i=0}^{2^n}\int_{x_i}^{x_{i+1}} f(x)\,dx \text{ with } h = (b-a)/2^n,\ x_i = a + ih, \qquad 1 \le i \le 2^n \\
&= \sum_{i=0}^{2^n}(x_{i+1} - x_i)f(x_i) + \frac{1}{2}\sum_{i=0}^{2^n}(x_{i+1} - x_i)^2 f'(\xi_i) \\
&= h\sum_{i=0}^{2^n} f(x_i) + \frac{h^2}{2}\sum_{i=0}^{2^n} f'(\xi_i) \\
&= h\sum_{i=0}^{2^n} f(a + ih) + E
\end{aligned}$$

where

$$\begin{aligned}
E &= \tfrac{1}{2}(b-a)h\left[\frac{1}{2^n}\sum_{i=0}^{2^n} f'(\xi_i)\right] \\
&= \tfrac{1}{2}(b-a)hf'(\xi) \text{ for } \xi \in (a,b)
\end{aligned}$$

5.1 Computer Exercises

5.1.C2. **a.** 2

b. 1.71828

c. 0.43882

5.1.C5. 1.8138

5.1.C9. 0.94598385, 0.94723395

5.1.C11. **a.** 4.84422

5.2 Romberg Algorithm

5.2.1. $R(5,3) = R(5,2) + \frac{1}{4^3-1}[R(5,2) - R(4,2)] = 12 + \frac{1}{63}[12 - -51)] = 13$

5.2.3.
$$\begin{aligned} R(4,1) &= R(4,0) + \frac{1}{4-1}[R(4,0) - R(3,0)] = 8 + \frac{1}{3}[8-8] = 8 \\ R(5,1) &= R(5,0) + \frac{1}{4-1}[R(5,0) - R(4,0)] = -4 + \frac{1}{3}[-4-8] = -8 \\ R(5,2) &= R(5,1) + \frac{1}{4^2-1}[R(5,1) - R(4,1)] = -8 + \frac{1}{15}[-8-8] = -\frac{136}{15} \approx -9.0667 \end{aligned}$$

5.2.5.
$$\begin{aligned} R(0,0) &= \frac{1}{2}(2-0)[4 + \frac{4}{1+2^2}] = \frac{24}{5} = 4.8 \\ R(1,0) &= \frac{1}{2}R(0,0) + \frac{1}{2}(b-a)f\left(\frac{a+b}{2}\right) = \frac{1}{2} \times \frac{24}{5} + \frac{1}{2}(2)\frac{4}{1+1} = \frac{22}{5} = 4.4 \\ R(1,1) &= \frac{4}{3}R(1,0) - \frac{1}{3}R(0,0) = \frac{4}{3}\left(\frac{22}{5}\right) - \frac{1}{3}\left(\frac{24}{5}\right) = \frac{64}{15} = 4.267 \end{aligned}$$

5.2.7. Method works but not well since integrand is not smooth. Romberg works on any continuous function.

5.2.8.
$$\begin{aligned} R(0,0) &= \frac{(2h)}{2}[f(-h) + f(h)] \\ R(1,0) &= \frac{h}{2}[f(-h) + 2f(0) + f(h)] \\ R(1,1) &= \frac{h}{3}\{f(-h) + 4f(0) + f(h)\} \quad \text{Simpson's rule} \end{aligned}$$

5.2.10. Only the first column in the Romberg array requires function evaluations. By (1)

n for $R(n+1,1)$	0	1	2	3	4	$\cdots$	$m-1$
# fcn. evaluations	2	1	2	2^2	2^3		2^{m-2}

Total # fcn. evaluations $= 2 + 1 + 2 + 2^2 + 2^3 + \cdots + 2^{m-2} = 1 + 2^{m-1}$ for $m \times m$ Romberg array. Note that $\sum_{k=0}^{n} 2^k = 2^{n+1} - 1$.

5.2.13. Let $h = (b-a)/4$. Hence, we have

$$R(1,1) = \frac{(2h)}{3}[f(a) + 4f(a+2h) + f(b)]$$
$$R(2,1) = \frac{h}{3}[f(a) + 4f(a+h) + 2f(a+2h) + 4f(a+3h) + f(b)]$$

Since $R(2,2) = \frac{16}{15}R(2,2) - \frac{1}{15}R(2,1)$, we have

$$R(2,2) = \frac{2h}{45}[7f(a) + 32f(a+h) + 12f(a+2h) + 32f(a+3h) + 7f(b)]$$

5.2.14. Since $X = S_{10} + C \cdot 10^{-3}$ and $X = S_{30} + C \cdot 30^{-3} = S_{30} + \frac{1}{27}C \cdot 10^{-3}$, we have

$$\begin{cases} u + C \cdot 10^{-3} = X \\ 27v + C \cdot 10^{-3} = 27X \end{cases}$$

Hence, we obtain

$$X = (27v - u)/26$$

5.2.15. Since $Z = f(h) + ah^2 + bh^4 + ch^6$, we have $Z = f(\frac{h}{2}) + \frac{1}{4}ah^2 + \frac{1}{16}bh^4 + \frac{1}{64}ch^6$.

Four times the latter minus the former gives $3Z = 4f(\frac{h}{2}) - f(h) - \frac{3}{4}bh^4 - \frac{15}{16}ch^6$.

Hence, we obtain $Z = \frac{4}{3}f(\frac{h}{2}) - \frac{1}{3}f(h) - \frac{1}{4}bh^4 - \frac{5}{16}ch^6$.

Now combine $\frac{4}{3}f(\frac{1}{2}) - \frac{1}{3}f(1)$, $\frac{4}{3}f(\frac{1}{4}) - \frac{1}{3}f(\frac{1}{2})$, $\frac{4}{3}f(\frac{1}{8}) - \frac{1}{3}f(\frac{1}{4})$, etc. to obtain improved estimates for Z.

Let $\varphi(h) = \frac{4}{3}f(\frac{h}{2}) - \frac{1}{3}f(h)$.

Since $Z = \varphi(h) - \frac{1}{2^2}bh^4 - \frac{5}{2^4}ch^6$, we have $Z = \varphi(\frac{h}{2}) - \frac{1}{2^6}bh^4 - \frac{5}{2^{10}}ch^6$.

Multiplying the latter by 2^4 and subtracting the former, we have $(2^4 - 1)Z = 2^4\varphi(\frac{h}{2}) - \varphi(h) + \frac{15}{2^6}ch^6$. Hence, we obtain $Z = [16\varphi(\frac{h}{2}) - \varphi(h)]/15 + \frac{1}{2^6}ch^6$.

Now let $\theta(h) = [16\varphi(\frac{h}{2}) - \varphi(h)]/15$.

Since $Z = \theta(h) + \frac{1}{2^6}ch^6$, we have $Z = \theta(\frac{h}{2}) + \frac{1}{2^{12}}ch^6$.

Multiplying the latter by 2^6 and subtracting the former, we have $(2^6 - 1)Z = 2^6\theta(\frac{h}{2}) - \theta(h)$.

Thus, we have $Z = [64\theta(\frac{h}{2}) - \theta(h)]/63$.

Hence, Z can be computed exactly from any three of the eleven numbers by

$$\begin{aligned} Z &= \frac{64}{63}\left(\frac{16}{15}\left[\frac{4}{3}f\left(\frac{h}{8}\right) - \frac{1}{3}f\left(\frac{h}{4}\right)\right] - \frac{1}{15}\left[\frac{4}{3}f\left(\frac{h}{4}\right) - \frac{1}{3}f\left(\frac{h}{2}\right)\right]\right) \\ &\quad - \frac{1}{63}\left(\frac{16}{15}\left[\frac{4}{3}f\left(\frac{h}{4}\right)\frac{1}{3}f\left(\frac{h}{2}\right)\right] - \frac{1}{15}\left[\frac{4}{3}f\left(\frac{h}{2}\right) - \frac{1}{3}f(h)\right]\right) \\ &= Af\left(\frac{h}{8}\right) + Bf\left(\frac{h}{4}\right) + Cf\left(\frac{h}{2}\right) + Df(h) \end{aligned}$$

where

$$A = \frac{4096}{2835}, \quad B = -\frac{1344}{2835}, \quad C = \frac{84}{2835}, \quad D = \frac{-1}{2835}$$

5.2.17. Since $L - x_n = a_3 n^{-3} + a_4 n^{-4} + \cdots$, we have

$$L - x_{n+1} = a_3(n+1)^{-3} + a_4(n+1)^{-4} + \cdots$$

Hence, we obtain

$$n(L - x_n) = a_3 + a_4 n^{-1} + \cdots$$

and

$$(n+1)^3(L - x_{n+1}) = a_3 + a_4(n+1)^{-1} + \cdots$$

Now subtracting, we have

$$[(n+1)^3 - n^3]L - [(n+1)^3 x_{n+1} - n^3 x_n] = a_4[(n+1)^{-1} - n^{-1}] + \cdots$$

Thus, we have

$$L = [(n+1)^3 x_{n+1} - n^3 x_n]/[(n+1)^3 - n^3] + \cdots$$

or

$$x_{n+1} + n^3(x_{n+1} - x_n)/(3n^2 + 3n + 1)$$

5.2.18. $I = R(n,0) + a_2 h^2 + a_4 h^4 + a_6 h^6 + \cdots$

$I = R(n+1,1) - \dfrac{1}{4} a_4 h^4 - \dfrac{5}{16} a_6 h^6$

$I = R(n+1,2) + \dfrac{1}{4^3} a_6 h^6 + \dfrac{21}{4^5} a_8 h^8$

In general, we obtain

$$|I - R(n,m)| = \mathcal{O}(h^{2m}) \text{ as } h \to 0$$

5.2.22. We have

$$\int_a^b f(x)\,dx = R(n,0) + a_3h^3 + a_6h^6 + \cdots$$

with $h = (b-a)/2^{n-1}$. Replacing h with $h/2$, we have

$$\int_a^b f(x)\,dx = R(n+1,0) + \frac{1}{8}a_3h^3 + \frac{1}{64}a_6h^6 + \cdots$$

Eliminating the a_3-term, we have

$$7\int_a^b f(x)\,dx = 8R(n+1,0) - R(n,0) - \frac{7}{8}a_6h^6 - \cdots$$

Let $R(n+1,1) = \frac{8}{7}\ R(n+1,0) - \frac{1}{7}\ R(n,0) = R(n+1,0)\ + [R(n+1,0) - R(n,0)]/7$. Continuing, we have

$$\int_a^b f(x)\,dx = R((n+1,1) - \frac{1}{8}a_6h^6 - \cdots$$

Replacing h with $\frac{h}{2}$ and n with $n+1$, we have

$$\int_a^b f(x)\,dx = R(n+2,1) - \frac{1}{8.64}a_6h^6 - \cdots$$

Eliminating the a_6-term, we have

$$63\int_a^b f(x)\,dx = 64R(n+2,1) - R(n+1,1) - \cdots$$

Let

$$R(n+2,2) = R(n+2,1) + [R(n+2,1) - R(n+1,1)]/63$$

Formula (2) becomes

$$R(n+1,m+1) = R(n+1,m) + [R(n+1,m) - R(n,m)]/(8^m - 1)$$

5.2.23. By (3), we have $\int_a^b f(x)\,dx - R(n-1,0) = a_2h^2 + \cdots$ where $h = (b-a)/2^{n-1}$.

Hence, the error is approximately $a_2[(b-a)2^{1-n}]^2 = a_2(b-a)^2 4^{1-n} = c4^{-n}$ where c is a constant.

5.2.24. Let $m = 1$ and let $n \to \infty$ in Formula (2).

5.2.27. $\displaystyle\int_0^{2\pi} (1 + \cos 4x)\, dx = x + \frac{1}{4}\sin 4x\Big|_0^{2\pi} = 2\pi$

$n = 5,\ h = \dfrac{2\pi}{4},\ x_i = (i-1)\dfrac{\pi}{2},\ f(x_i) = 1 + \cos((i-1)2\pi)$

$$\begin{aligned}\frac{h}{2}\sum_{i=1}^{n-1}[f(x_i) + f(x_{i+1})] &= \frac{h}{2}[f(x_1) + 2f(x_2) + 2f(x_3) + 2f(x_4) + f(x_5)]\\ &= \frac{\pi}{4}(2 + 4 + 4 + 4 + 2) = 4\pi\end{aligned}$$

$E = 2\pi - 4\pi = -2\pi$ (true Error)

$$E = \sum_{k=1}^{m-1} A_{2k}h^{2k}[f^{(2k-1)}(a) - f^{(2k-1)}(b)] + A_{2m}(b-a)h^{2m}f^{(2m)}(\xi)$$

Now
$$\begin{aligned} f(x) &= 1 + \cos 4x\\ f'(x) &= -4\sin 4x\\ f''(x) &= -4^2\cos 4x\\ f'''(x) &= 4^3\sin 4x\\ f^{(4)}(x) &= 4^4\cos 4x\\ f^{(5)}(x) &= -4^5\sin 4x\\ f^{(6)}(x) &= -4^6\cos 4x\end{aligned}$$

Hence, we obtain

$$E = A_{2m}(2\pi)\left(\frac{2\pi}{4}\right)^{2m}[\pm 4^{2m}\cos(4\xi)] \pm (2\pi)^{2m+1}4^{2m+1}A_{2m}\cos(4\xi)$$

5.2 Computer Exercises

5.2.C1. $R(7,7) = 0.49996\ 9819$

5.2.C5. $R(5,0) = 1.81379\ 9364$

5.2.C6. $2/9 = 0.22222\ldots$

5.2.C7. $0.62135\ 732$

5.2.C11. $R(7,7) = 0.76519\ 7687$

5.3 Simpson's Rule and Newton-Cote Rules

5.3.1. $\int_0^1 f(x)\,dx \approx \frac{h}{3}\{f(0) + 4f(\frac{1}{2}) + f(1)\}$ where $h = \frac{1}{2}$ and $f(x) = \frac{1}{(1+x^2)}$. Now

$$\text{RHS} = \frac{1}{6}\left\{1 + 4(\frac{4}{5}) + \frac{1}{2}\right\} = \frac{47}{60} = 0.7833$$

The true solution is

$$\text{LHS} = \arctan x\Big|_0^1 = \frac{\pi}{4} = 0.7854$$

and the error is 0.23×10^{-2}.

5.3.2. **a.** The error term for the trapezoid rule is $-\frac{(b-a)}{12}h^2 f''(\xi)$ for $n-1$ subintervals.
Now $f(x) = \sin\left(\frac{\pi}{2}x^2\right)$, $f'(x) = \pi x \cos\left(\frac{\pi}{2}x^2\right)$,
$f''(x) = \pi \cos\left(\frac{\pi}{2}x^2\right) - (\pi x)^2 \sin\left(\frac{\pi}{2}x^2\right)$ and $|f''(x)| \le \pi + \pi^2$ over the interval $[0,1]$.
We require | error term | $\le \frac{1}{12}h^2(1+\pi)\pi < 10^{-3}$ or $h^2 < \frac{12 \times 10^{-3}}{\pi(1+\pi)} \approx 9.2 \times 10^{-4}$.
Hence, we have $h < 0.03$ or $n > 33.97$. So let $n = 34$ or $h = 0.0294$.

b. The error term for Simpson's $\frac{1}{3}$-rule is $-\frac{(b-a)h^4}{180} f^{(4)}(\xi)$ for $n-1$ subintervals.
Now

$$\begin{aligned} f'''(x) &= -3\pi^2 x \sin\left(\tfrac{\pi}{2}x^2\right) - (\pi x)^3 \cos\left(\tfrac{\pi}{2}x^2\right) \\ f^{(4)}(x) &= -3\pi^2 \sin\left(\tfrac{\pi}{2}x^2\right) - 3\pi^3 x^2 \cos\left(\tfrac{\pi}{2}x^2\right) \\ &\quad -3\pi^3 x^2 \cos\left(\tfrac{\pi}{2}x^2\right) + (\pi x)^4 \sin\left(\tfrac{\pi}{2}x^2\right) \end{aligned}$$

and $|f^{(4)}(x)| \le 3\pi^2 + 6\pi^3 + \pi^4 = \pi^2(3 + 6\pi + \pi^2) \approx 313.1$ over the interval $[0,1]$.
We require | Error term | $\le \frac{h^4}{180}(313.1) < 10^{-3}$ or $h^4 < \frac{0.180}{313.1} = 5.75 \times 10^{-4}$.
Hence, we obtain $h < 0.15$ or $n > 7.5$.
We take $n = 9$ or $h = 0.125$.

5.3.3. **a.** $\displaystyle\int_1^2 f(x)\,dx \approx \frac{0.5}{3}[10+4\cdot 7+5] = \frac{43}{6} = 7.1667.$

b. $\displaystyle\int_1^2 f(x)\,dx \approx \frac{0.25}{3}[10+4\cdot 8+2\cdot 7+4\cdot 6+5] = \frac{85}{12} = 7.0833.$

c. $I \approx 7.1667 + Ch^4$

$I \approx 7.0833 + C(h/2)^4$

Thus, we have $15I \approx 16(7.0833) - (7.1667)$ or $I = 7.0777$.

5.3.4. Let $h = 0.25$ and

$$\int_1^2 \frac{dx}{x} = \frac{0.25}{3}[1+4(1.25)^{-1}+2(1.5)^{-1}+4(1.75)^{-1}+(2)^{-1}] = 0.6933$$

Error $= -\dfrac{(b-a)}{180}h^4 f^{(4)}$.

Now $f = x^{-1}$, $f' = -x^{-2}$, $f'' = 2x^{-3}$, $f''' = -6x^{-4}$, and $f^{(4)} = 24x^{-5}$.

Hence, we obtain $|f^{(4)}| \le 24$ and $|$ Error $| \le \dfrac{(.25)^4(24)}{(180)} = 5.2\times 10^{-4}$ is our bound.

Notice that $\ln 2 = 0.693147$ so that the error is really 1.5×10^{-4} which is less than the bound.

5.3.7. We are given $\displaystyle\int_a^b f(x)\,dx = S_{n-1} + Ch^4$. Consider with $2(n-1)$ so h is replaced by $h/2$,

$$\int_a^b f(x)\,dx = S_{2(n-1)} + \frac{Ch^4}{16}$$

Eliminating the Ch^4 terms, we have

$$15\int_a^b f(x)\,dx = 16S_{2(n-1)} - S_{ny-1}$$

so that

$$\int_a^b f(x)\,dx = \frac{16}{15}S_{2(n-1)} - \frac{1}{15}S_{n-1}$$

5.3.8. By Taylor series, we have

$$f(a+h) = f + hf' + \frac{1}{2}h^2 f'' + \frac{1}{6}h^3 f''' + \frac{1}{4!}h^4 f^{(4)} + \cdots$$
$$f(a+2h) = f + 2hf' + 2h^2 f'' + \frac{4}{3}h^3 f''' + \frac{2^4}{4!}h^4 f^{(4)} + \cdots$$
$$f(a+3h) = f + 3hf' + \frac{9}{2}h^2 f'' + \frac{9}{2}f''' + \frac{3^4}{4!}h^4 f^{(4)} + \cdots$$

Hence, we obtain

$$f + 3f(a+h) + 3f(a+2h) + f(a+3h) = 8f + 12hf' + 12h^2 f'' + 9h^3 f''' + \frac{11}{2}h^4 f^{(4)} + \cdots$$

and

$$\frac{3h}{8}[f + 3f(a+h) + 3f(a+2h) + f(a+3h)] = 3hf + \frac{9}{2}h^2 f' + \frac{9}{2}h^3 f'' + \frac{27}{8}h^4 f''' + \frac{33}{16}h^5 f^{(4)} + \cdots$$

Letting $3h$ replace h in (2) of Section 4.2, we have

$$\int_a^{a+3h} f(x)\,dx = 3hf + \frac{9}{2}h^2 f' + \frac{9}{2}h^3 f'' + \frac{27}{8}h^4 f''' + \frac{81}{40}h^5 f^{(4)} + \cdots$$

Consequently,

$$\int_a^{a+3h} f(x)\,dx = \frac{3h}{8}[f(a) + 3f(a+h) + 3f(a+2h) + f(a+3h)] - \frac{3}{80}h^5 f^{(4)} + \cdots$$

The error term is $-\frac{3}{80}h^5 f^{(4)}(\xi)$ for some ξ between a and $a+3h$.

Both Simpson's $\frac{1}{3}$-rule and $\frac{3}{8}$-rule have error terms $\mathcal{O}(h^5)$.

Since the Simpson's $\frac{3}{8}$-rule requires one more function evaluation than the Simpson's $\frac{1}{3}$-rule for an approximation of the same order of accuracy, it is clearly less desirable.

5.3 Computer Exercises

5.3.C1. 3.1416; 3.1416

5.3.C2. 0.1996

5.4 Gaussian Quadrature Formulas

5.4.1. $a = 0$, $b = 2$, $f(x) = e^{-x^2}$

$$\int_0^2 \exp(-x^2)\,dx \approx \exp\left[-\left(-\sqrt{\frac{1}{3}}+1\right)^2\right] + \exp\left[-\left(\sqrt{\frac{1}{3}}+1\right)^2\right] \approx 0.91949$$

5.4.4. **a.** $n = 2$ $\quad q_2(x) = \frac{3}{2}x^2 - \frac{1}{2} = 0 \quad x = \pm\sqrt{\frac{1}{3}}$

b. $n = 4$

$$\begin{aligned} q_4(x) &= \frac{7}{4}xq_3(x) - \frac{3}{4}q_2(x) \\ &= \frac{35}{8}x^4 - \frac{21}{8}x^2 - \frac{9}{8}x^2 + \frac{3}{8} \\ &= \frac{35}{8}x^4 - \frac{30}{8}x^2 + \frac{3}{8} \end{aligned}$$

Solve $35y^2 - 30y + 3 = 0$ where $y = x^2$. Hence, we obtain

$$\begin{aligned} y &= \frac{30 \pm \sqrt{900-420}}{70} = \frac{3}{7} \pm \frac{4}{7}\sqrt{\frac{3}{10}} \\ x &= \pm\sqrt{\frac{1}{7}(3 \pm 4\sqrt{0.3})} \\ x &= \pm 0.861136, \quad \pm 0.339981 \end{aligned}$$

5.4.5. Let LHS $= \int_{-1}^{1} f(x)\,dx$ and RHS $= \alpha f(-\frac{1}{2}) + \beta f(0) + \gamma f(\frac{1}{2})$.

$f(x)$	LHS	RHS
1	2	$= \alpha + \beta + \gamma$
x	0	$= -\frac{1}{2}\alpha + \frac{1}{2}\gamma$
x^2	$\frac{2}{3}$	$= \frac{1}{4}\alpha + \frac{1}{4}\gamma$

Hence, for LHS = RHS, we have $\alpha = \gamma = \frac{4}{3}$, $\beta = -\frac{2}{3}$.

5.4.6. Let LHS $= \int_a^b f(x)\,ds$ and RHS $= Af(a) + Bf'(b)$.

$f(x)$	LHS	RHS
1	$b - a$	$= A$
$x - a$	$\frac{1}{2}(b-a)^2$	$= B$

Hence, we obtain $A = (b - a)$, $B = \frac{1}{2}(b-a)^2$.

5.4.7. Let LHS $= \int_a^{a+h} f(x)\,dx$ and RHS $= \alpha f(a) + \beta f(a+h) + \gamma f(a+2h)$.

$f(x)$	LHS		RHS
1	h	$=$	$\alpha+\beta+\gamma$
$x-a$	$\frac{h^2}{2}$	$=$	$\beta h + \gamma 2h$
$(x-a)^2$	$\frac{h^3}{3}$	$=$	$\beta h^2 + \gamma(2h)^2$

Letting LHS = RHS, we have

$$\begin{cases} \alpha+\beta+\gamma = h \\ \beta + 2\gamma = \dfrac{h}{2} \\ \beta + 4\gamma = \dfrac{h}{3} \end{cases}$$

Hence, we obtain $\beta = \frac{2h}{3}$, $\gamma = -\frac{h}{12}$, $\alpha = \frac{5h}{12}$.

5.4.9. Let LHS $= \int_{-1}^{1} f(x)x^2\,dx$ and RHS $= af(-\alpha) + bf(0) + cf(\alpha)$.

$f(x)$	LHS		RHS
1	$\frac{2}{3}$	$=$	$a+b+c$
x	0	$=$	$-a\alpha + c\alpha$
x^2	$\frac{2}{5}$	$=$	$a\alpha^2 + c\alpha^2$
x^3	0	$=$	$-a\alpha^3 + c\alpha^3$
x^4	$\frac{2}{7}$	$=$	$a\alpha^4 + c\alpha^4$

Letting LHS = RHS, we have

$$\begin{cases} a + b + c = \dfrac{2}{3} \\ a \qquad - c = 0 \\ a \qquad + c = \dfrac{2}{5}\alpha^{-2} \\ a \qquad + c = \dfrac{2}{7}\alpha^{-4} \end{cases}$$

Hence, we obtain $\alpha = \sqrt{5/7}$, $a = c = 7/25$, and $b = 8/75$.

5.4.10. Let LHS $= \int_0^h f(x)\,dx$ and RHS $= w_1 f(0) + w_2 f(h) + w_3 f''(0) + w_4 f''(h)$

$f(x)$	$f''(x)$	LHS		RHS
1	0	h	$=$	$w_1 + w_2$
x	0	$\frac{h^2}{2}$	$=$	$w_2 h$
x^2	2	$\frac{h^3}{3}$	$=$	$w_2 h^2 + 2w_3 + 2w_4$
x^3	$6x$	$\frac{h^4}{4}$	$=$	$w_2 h^3 \qquad + 6w_4 h$

Setting LHS = RHS, we have $w_2 = \frac{h}{2}$, $w_1 = \frac{h}{2}$, $w_4 = -\frac{h^3}{24}$, $w_3 = -\frac{h^3}{24}$.

5.4.11. Let LHS $= \int_{-h}^h f(x)\,dx$ and RHS $= Af(0) + Bf'(-h) + Cf''(h)$.

$f(x)$	$f'(x)$	$f''(x)$	LHS RHS
1	0	0	$2h = A$
x	1	0	$0 = B$
x^2	$2x$	2	$\frac{2}{3}h^3 = -2Bh + 2C$

Hence, we obtain $A = 2h$, $B = 0$, $C = \frac{h^3}{3}$.

5.4.12. LHS $= \int_{-2}^{+2} f(x)\,dx \quad$ RHS $= Af(-1) + Bf(0) - Cf(+1)$

f	LHS		RHS	
1	4	$=$	$A + B + C$	
x	0	$=$	$-A$	$+C$
x^2	$\frac{16}{3}$	$=$	A	$+C$
x^3	0	$=$	$-A$	$+C$
x^4	$\frac{64}{4}$	$\neq$	A	$+C$

$$A = \frac{8}{3}, \quad B = -\frac{4}{3}, \quad C = \frac{8}{3}$$

Yes. Exact for polynomials of degree ≤ 3.

5.4.13. Let LHS $= \int_{-h}^{h} f(x)\,dx$, RHS $= Af(-h) + Bf(0) + Cf(h) - hDf'(h)$

f	f'	LHS		RHS
1	0	$2h$	$=$	$A + B + C$
x	1	0	$=$	$-hA + hC - hD$
x^2	$2x$	$\frac{2}{3}h^3$	$=$	$h^2A + h^2C - 2h^2D$
x^3	$3x^2$	0	$=$	$-h^3A + h^3C - 3h^3D$

$$\begin{cases} A + B + C = 2h \\ A - C + D = 0 \\ A + C - 2D = \frac{2}{3}h \\ A - C + 3D = 0 \end{cases}$$

$$\begin{cases} 2A - D = \frac{2}{3}h \\ 2A + D = \frac{2}{3}h \end{cases}$$

$$\begin{cases} A = \frac{h}{3} \\ B = \frac{4}{3}h \\ C = \frac{h}{3} \\ D = 0 \end{cases}$$

5.4.14. LHS $= \int_0^{3h} f(x)\,dx$, RHS $= \frac{3h}{8}\{f(0) + 3f(h) + 3f(2h) + f(3h)\}$

f	LHS		RHS
1	$3h$	$=$	$\frac{3h}{8}\{1 + 3 + 3 + 1\}$
x	$\frac{9h^2}{2}$	$=$	$\frac{3h}{8}\{0 + 3h + 6h + 3h\}$
x^2	$\frac{27h^3}{3}$	$=$	$\frac{3h}{8}\{0 + 3h^2 + 12h^2 + 9h^2\}$
x^3	$\frac{81h^4}{4}$	$=$	$\frac{3h}{8}\{0 + 3h^3 + 24h^3 + 27h^3\}$
x^4	$\frac{243h^5}{5}$	$\neq$	$\frac{3h}{8}\{0 + 3h^4 + 48h^4 + 81h^4\}$

True for $n \leq 3$.

5.4 Computer Exercises

5.4.C2. **a.** 1.4183

5.4.C8. **a.** 2.03480 53185 7741

b. 0.89297 95115 6925

c. 0.43398 77196 7972

The MATLAB results using quad is very far from the correct value 0.42682 08882 3553 obtained in Computer Exercise 5.4.9.

Chapter 6

Spline Functions

6.1 First-Degree and Second-Degree Splines

6.1.1. The function is a first-degree spline.

It is only necessary to verify its continuity at the knots: $S_0(1/2) = 0.5 = S_1(1/2)$ and $S_1(2) = 3.5 = S_1(2)$.

6.1.6. Assume formula Exercise 6.1.5 is correct. We have

$$\begin{aligned}
|f(x) - p(x)| &\leq \frac{|x-a||x-b|}{|b-a|}\left|\frac{f(x)-f(b)}{x-b} - \frac{f(x)-f(a)}{x-a}\right| \\
&\leq \frac{|x-a||x-b|}{|b-a|}|f'(\xi_1)| + |f(\xi_2)| \\
&\leq \frac{|x-a||x-b|}{|b-a|}2C \\
&\leq \left[\left(\tfrac{1}{2}(b-a)\right)^2/(b-a)\right][2C] \\
&= \tfrac{1}{4}(b-a)(2C) = \tfrac{1}{2}(b-a)C = \tfrac{1}{2}\ell C
\end{aligned}$$

In Exercise 6.1.5, the bracketed expression is $f'(\xi_1) - f'(\xi_2)$ and in magnitude does not exceed $2C$.

Also, the following calculation is needed: Let $a \leq x \leq b$.

Then $(x-a)(b-a) = xb - x^2 - ab + ax$.

Derivative is $a + b - 2x$ and root is $(a+b)/2$.

Value of function here is
$[\frac{1}{2}(a+b) - a][b - \frac{1}{2}(a+b)] = [\frac{1}{2}(b-a)][\frac{1}{2}(b-a)] = [\frac{1}{2}(b-a)]^2 = \frac{1}{4}\ell^2$.

6.1.9. $f(x) = \sin 100x, \quad f'(x) = 100\cos 100x, \quad |f'(x)| \le 100.$

Use Exercise 9.1.7, $|f(x) - S(x)| \le \frac{1}{2}C\delta \equiv \frac{1}{2} \cdot 100 \cdot \delta < 10^{-8}$.

So $\frac{1}{2}10^2\delta < 10^{-8}$ and $\delta < 2 \cdot 10^{-10}$.

If n knots are equally spaced on $[0, \pi]$, there are n subintervals of width π/n.

So we want $\dfrac{\pi}{n} < 2 \times 10^{-10}$.

Here $\dfrac{n}{\pi} > \frac{1}{2} \cdot 10^{10}$ and $n > \dfrac{\pi}{2} \cdot 10^{10} \quad \Rightarrow \quad n \approx 1.57 \times 10^{10}$.

6.1.10. $\sum_{i=0}^{n} f(t_i)S_i$ is a linear combination of 1st-degree spline functions having knots $t_0, \ldots, t_n$.

Hence, it is also such a function.

Its value at t_j is $\sum_{i=0}^{n} f(t_i)S_i(t_j) = f(t_j)$. $S_i(x) = 0$ if $x < t_{i-1}$ or $x > t_{i+1}$.

On (t_{i-1}, t_i), $S_i(x)$ is given by $(x - t_{i-1})/(t_i - t_{i-1})$.

On (t_i, t_{i+1}), $S_i(x)$ is given by $(x - t_{i+1})/(t_i - t_{i+1})$. S_0 and S_n are slightly different.

6.1.12. If S is piecewise quadratic, then clearly S' is piecewise linear.

If S is a quadratic spline then $S \in C^1$.

Hence, we obtain $S' \in C$.

Hence, S' is piecewise linear and continuous.

6.1.17.

i	1	2	3	4	5	6
t_i	-1	0	1/2	1	2	5/2
y_i	2	1	0	1	2	3
h_i	1	1/2	1/2	1	1/2	
z_i	0	-2	-2	6	-4	8

$$\begin{aligned}
Q_0(x) &= -(x+1)^2 + 2 \\
Q_1(x) &= -2x + 1 \\
Q_2(x) &= 8(x - \tfrac{1}{2})^2 - 2(x - \tfrac{1}{2}) \\
Q_3(x) &= -5(x-1)^2 + 6(x-1) + 1 \\
Q_4(x) &= 12(x-2)^2 - 4(x-2) + 2
\end{aligned}$$

6.1.19. The answer is given by Equation (8).

6.1.20.

a. Yes. $Q(x)$ is piecewise quadratic, and $Q(x)$ and $Q'(x)$ are continuous at 1.

b. No. $Q'(x)$ is not continuous at $x = 0$ and $Q_0''(-100) \neq 0$.

c. No. $Q'(x)$ is not continuous at $x = 1, 2$.

6.1.21. Yes.

6.2 Natural Cubic Splines

6.2.1. $S(x) = \begin{cases} ax^3 + x^2 + cx, & -1 \le x \le 0 \\ bx^3 + x^2 + dx, & 0 \le x \le 1 \end{cases}$

$x = -1 : -a + 1 - c = 1 \qquad a + c = 0 \qquad (1)$

$x = 1 : \quad b + 1 + d = 1 \qquad b + d = 0 \qquad (2)$

$S'(x) = \begin{cases} 3a^2 + 2x + c, & -1 \le x \le 0 \\ 3bx^2 + 2x + d, & 0 \le x \le 1 \end{cases}$

$S'(x)$ continuous at $x = 0 \Rightarrow c = d \qquad (3)$

$S''(x) = \begin{cases} 6ax + 2, & -1 \le x \le 0 \\ 6bx + 2, & 0 \le x \le 1 \end{cases}$

$S''(-1) = S''(1) = 0$

$\Rightarrow \begin{cases} -6a + 2 = 0, & a = \frac{1}{3} \qquad (4) \\ 6b + 2 = 0, & b = -\frac{1}{3} \qquad (5) \end{cases}$

(1) & $(4) \Rightarrow c = -\frac{1}{3}$

(2) & $(5) \Rightarrow d = \frac{1}{3}$

From this $c = -d$ but from (3) $c = d$ and we have a contradiction! Therefore, there is no such function.

6.2.2. $S_0(-1) = S_1(-1) \Rightarrow 1 = -a + b - c + d \quad \Rightarrow \quad \begin{cases} b + d = 1 \\ a + c = 0 \end{cases}$

$S_1(1) \ = S_2(1) \quad \Rightarrow 1 = a + b + c + d$

$S_0'(-1) = S_1'(-1) \ -1 = 3a - 2b + c \qquad \Rightarrow \quad \begin{cases} 3a + c = 0 \\ b = \frac{1}{2} \end{cases}$

$S_1'(1) \ = S_2'(1) \qquad 1 = 3a + 2b + c$

So $a = 0,\ c = 0,\ b = \frac{1}{2},\ d = \frac{1}{2}$.

$S_0''(-1) = S_1''(-1) \Rightarrow 0 = -6a + 2b \qquad \Rightarrow \begin{cases} b = 0 \\ a = 0 \end{cases}$

$S_1''(1) \ = S_2''(1) \quad \Rightarrow 0 = 6a + 2b$

$S''(-10) = S'(10) = 0$

Contradiction $b = \frac{1}{2}$ and $b = 0$!

Cannot be a natural cubic spline.

6.2.4. $a = -4;\ b = -6;\ c = -3;\ d = -1;\ e = -3$

6.2.5. $f(x) = \begin{cases} 3 + x - 9x^3, & [0,1] \\ a + b(x-1) + c(x-1)^2 + d(x-1)^3, & [1,2] \end{cases}$

$f'(x) = \begin{cases} 1 - 27x^2, & [0,1] \\ b + 2c(x-1) + 3d(x-1)^2, & [1,2] \end{cases}$

$f''(x) = \begin{cases} -54x, & [0,1] \\ 2c + 6d(x-1), & [1,2] \end{cases}$

Continuity of f at $x = 1$ gives us $a = -5$

Continuity of f' at $x = 1$ gives us $b = -26$

Continuity of f'' at $x = 1$ gives us $c = -27$

So far, $f(x) = -5 - 26(x-1) - 27(x-1)^2 + d(x-1)^3$.

To minimize $\int_0^2 (f'')^2$ we only have to minimize $\int_1^2 (f'')^2$ because f'' is fixed on $[0,1]$.

So we minimize $\int_1^2 [-54 + 6d(x-1)]^2\, dx$.

Change variable $x = t + 1$. $x = 1 \Rightarrow t = 0$, $x = 2 \Rightarrow t = 1$,

$dx = dt$ minimize $\int_0^1 (-54 + 6d \cdot t)^2\, dt$.

The derivative of this with respect to d is taken and set equal to 0:

$\int_0^1 2(-54 + 6d \cdot t)(6t)\, dt = 0$ Cancelling factor of $2 \cdot 6 \cdot 6$.

We have $\int_0^1 -9t\, dt + d \int_0^1 t^2\, dt = 0$ or $d = 27/2$.

6.2.6. No. $f(x)$ is not continuous at $x = 0$.

6.2.7. **a.** $S(x)$ is not continuous at $x = -1$
$S''(x)$ is not continuous at $x = -1, 1$

6.2.8. **a.** $(m+1)n$

b. $2n$

c. $(m-1)(n-1)$

d. $m - 1$

6.2.10. $S_0 = a + bx + cx^2$

$S_0' = b + 2cx$

$S_0'' = 2c$

$S_0(1) = a + b + c$, $S_1'(1) = b + 2c$, $S_1''(1) = 2c$

$S_1 = (a + b + c) + (b + 2c)(x - 1) + c(x - 1)^2 + d(x - 1)^3$

$S_1' = b + 2c + 2c(x - 1) + 3d(x - 1)^2$

$S_1'' = 2c + 6d(x - 1)$

$S_1(2) = a + b + c + b + 2c + c + d = a + 2b + 4c + d$

$S_1'(2) = b + 2c + 2c + 3d = b + 4c + 3d$

$S_1''(2) = 2c + 6d$

$S_2 = (a + 2b + 4c + d) + (b + 4c + 3d)(x - 2) + (c + 3d)(x - 2)^2$

Simple case:

$$S = \begin{cases} x^2, & [0,1] \\ 1 + 2(x-1) + (x-1)^2 + (x-1)^3, & [1,2] \\ 5 + 7(x-2) + 4(x-2)^2, & [2,3] \end{cases}$$

6.2.12. Denote the cubics by $S_0(x)$ and $S_1(x)$.

The continuity condition $S_0(1) = S_1(1)$ forces us to take $c = 1$.

The condition $S_0'(1) = S_1'(1)$ forces us to set $b = 3$.

The condition $S_0''(1) = S_1''(1)$ forces us to put $a = 3$.

6.2.13. Denote the parts of the function by S_0, S_1, and S_2.

Then $S_0'' = S_1'' = 0$, so $S_0''(-1) = S_1''(-1) = 0$ and $S_2''(1) = S_1''(1) = 0$.

But S_1 is cubic, and S_1'' is linear.

If S_1'' vanishes at -1 and $+1$ it must be 0.

Then S_1 itself is linear.

Hence, S is piecewise linear.

Continuity in S' necessarily fails.

6.2.15. If we impose continuity of S, S', and S'' at $x = 1$ we are led to the equations

$$\begin{cases} a + b + \ \ c + \ \ d = 2 \\ \quad b + 2c + 3d = 5 \\ \quad\quad 2c + 6d = 8 \end{cases}$$

The condition $S_2'''(x) = 12$ requires us to take $d = 2$.

Then $c = -2$, $b = 3$, and $a = -1$.

6.2.17. At each interior knot, the continuity of S, S', S'', S''' must be imposed.

This yields $4(n-2)$ conditions.

There are $5(n-1)$ coefficients.

Hence, $n+3$ additional conditions are needed to specify the spline completely.

Notice the statement of the exercise does not include any interpolation conditions.

6.2.19. $f'(x) = 3x^2 + 1$ on $(-\infty, 0)$ and $f'(x) = 3x^2 - 1$ on $(0, \infty)$.

Hence, f' is discontinuous at 0.

Hence, f is not a cubic spline.

Note that $f''(0)$ does not exist.

6.2.22. Let $p(x) = a + bx + cx^2 + dx^2$.

Since $p(0) = 0$ and $p'(0) = 1$ we have

$$\begin{cases} p(x) = x + cx^2 + dx^3 \\ p'(x) = 1 + 2cx + 3dx^2 \end{cases}$$

Substituting $x = 1$, we have

$$\begin{cases} 1 + c + d = 1.1752 \\ 1 + 2c + 3d = 1.5431 \end{cases}$$

The solution is then $p(x) = x - 0.0175x^2 + 0.1927x^3$.

This is not a natural cubic spline.

6.2.26. S is linear on $[t_1, t_2]$.

A cubic spline has a continuous second derivative.

Since S is linear on $[t_0, t_1]$ we have $S'' = 0$ on $[t_0, t_1]$.

In particular, $S''(t_1) = 0$.

Similarly, $S''(t_2) = 0$.

But on $[t_1, t_2]$, S is cubic and S'' is linear.

Since S'' vanishes at t_1 and t_2 we have $S'' \equiv 0$ on $[t_1, t_2]$.

Hence, S is linear on $[t_1, t_2]$.

Note that on $[t_0, t_3]$ S is a single linear polynomial.

6.2.32. If the algorithm is used, the values obtained are

$$\begin{cases} h_0 = h_1 = h_2 = h_3 = 1 \\ b_0 = 1,\ b_1 = -1,\ b_2 = 1,\ b_3 = -1 \\ u_1 = 4,\ u_2 = 15/4,\ u_3 = 56/15 \\ v_1 = -12,\ v_2 = 15,\ v_3 = -16 \\ z_4 = 0,\ z_3 = -30/7,\ z_2 = 36/7,\ z_1 = -30/7,\ z_0 = 0 \end{cases}$$

Then the component cubic functions are

$$\begin{aligned} S_0(x) &= (-5/7)(x-1)^3 + (12/7)(x-1) \\ S_1(x) &= (6/7)(x-2)^3 - (5/7)(3-x)^3 - (6/7)(x-2) + (12/7)(3-x) \\ S_2(x) &= (-5/7)(x-3)^3 + (6/7)(4-x)^3 + (12/7)(x-3) - (6/7)(4-x) \\ S_3(x) &= (-5/7)(5-x)^3 + (12/7)(5-x) \end{aligned}$$

6.2.33. The conditions on S make it an even function.

If $S(x) = S_0(x)$ in $[-1, 0]$ and $S(x) = S_1(x)$ in $[0, 1]$, then $S_1(0) = 1$, $S_1'(0) = 0$, $S_1''(1) = 0$, and $S_1(1) = 0$.

An easy calculation yields $S_1(x) = 1 - \frac{3}{2}x^2 + \frac{1}{2}x^3$.

6.2.38. $n+1$ knots define n subintervals.

So $5n$ coefficients are needed to define a piecewise quartic function.

The function would be a quartic spline if the function, its first, second, and third derivative are continuous at $(n-1)$ interior points:

These conditions give $4(n-1)$ conditions.

We need to impose $5n - 4(n-1) = n + 4$ more conditions.

6.2.39. Yes.

6.3 B-Splines: Interpolation and Approximation

6.3.2. Chebyshev polynomials recurrence relation. See Chapter 9 (p. 436).

6.3.3.

$$B_i^2(x) = \begin{cases} \dfrac{(x-t_i)^2}{(t_{i+2}-t_i)(t_{i+1}-t_i)}, & \text{on } [t_i, t_{i+1}] \\ \dfrac{(x-t_i)(t_{i+2}-x)}{(t_{i+2}-t_i)(t_{i+2}-t_{i+1})} + \dfrac{(t_{i+3}-x)(x-t_{i+1})}{(t_{i+3}-t_{i+1})\,,(t_{i+2}-t_{i+1})}, & \text{on } [t_{i+1}, t_{i+2}] \\ \dfrac{(t_{i+3}-x)^2}{(t_{i+3}-t_{i+1})(t_{i+3}-t_{i+2})}, & \text{on } [t_{i+2}, t_{i+3}] \\ 0, & \text{elsewhere} \end{cases}$$

6.3.5. A linear combination of piecewise linear continuous functions is again piecewise linear and continuous, hence a first-degree spline.

Since $B_i^1(t_{i+1}) = 1$, but $B_i^1(t_j) = 0$ for $j \neq i+1$, we have $\sum_{i=-\infty}^{\infty} f(t_i)B_{i-1}^1(t_j) = \sum_{i=-\infty}^{\infty} f(t_i)\delta_{ij} = f(t_j)$.

The 0-degree spline with the interpolation property is $\sum_{i=-\infty}^{\infty} f(t_i)B_i^0(x)$.

6.3.14. It was established in Section 6.3 that the support of B_i^k is the interval (t_i, t_{i+k+1}).

In order that this interval intersect the interval (t_n, t_m), it is necessary and sufficient that $t_n \in (t_i, t_{i+k+1})$ or $t_m \in (t_i, t_{i+k+1})$.

An equivalent condition is $i < n < i+k+1$ or $i < m < i+k+1$.

This occurs for exactly those indices i which satisfy $n-k \leq i \leq m-1$.

6.3.15. Use induction.

In Section 6.3, a formula for B_i^1 is given.

It shows that $B_i^1(x) = (x-t_i)/(t_{i+1}-t_i)$ on the interval $[t_i, t_{i+1}]$.

If the formula in the exercise is true for index $k-1$, then on $[t_i, t_{i+1}]$ we have, by Equation (3),

$$B_i^k(x) = \frac{x-t_i}{t_{i+k}-t_i}B_i^{k-1}(x) = \frac{x-t_i}{t_{i+k}-t_i}\,\frac{(x-t_i)^{k-1}}{(t_{i+1}-t_i)\ldots(t_{i+k+1}-t_i)}$$

Here we have used the fact that $B_{i+1}^{k+1}(x) = 0$ on $[t_i, t_{i+1}]$.

6.3.16. In Section 6.3, it was shown that if $S(x) = \sum_{i=-\infty}^{\infty} c_i B_i^0(x)$ then $S(t_j) = c_j$.

If $S(t_j)$ is specified for one or more values of j, then the corresponding c_j are fixed.

But the remaining c_j are arbitrary.

So there is no uniqueness.

However, if all remaining c's are set equal to zero, this additional condition specifies S uniquely.

Similar remarks apply to $\sum_{i=-\infty}^{\infty} c_i B_i^1(x)$.

For $k \geq 2$ the situation is like $k = 2$, as described in Section 6.3 (no uniqueness).

6.3.17. In Section 6.3, it was shown how to obtain a spline $\sum_{i=-\infty}^{\infty} A_i B_{i-2}^2(x)$ which takes prescribed values at the knots.

If these values are zero, then by Equation (4),

$$A_j(t_{j+1} - t_j) + A_{j+1}(t_j - t_{j-1}) = 0 \quad (-\infty < j < \infty)$$

There can be nontrivial solutions.

For example, let $A_0 = 1$ and use the above equation to compute all other A's recursively.

For $k = 0$ and 1, if $\sum_{i=-\infty}^{\infty} c_i B_i^k(x)$ vanishes at each knot, then all coefficients are zero, as shown in Section 6.3.

A concrete example is obtained by letting $t_i = i$ and $A_i = (-1)^i$ for $-\infty < i < \infty$.

Then $\sum_{i=-\infty}^{\infty} A_i B_i^2(t_j) = 0$ for all j.

6.3.19. Put $S(x) = \sum_{i=-\infty}^{\infty} c_i B_i^1(x)$.

If $S(x) = x$, then $S(t_j) = t_j$ and so $\sum_{i=-\infty}^{\infty} c_i B_i^1(t_j) = t_j$.

Since $B_i^1(t_j) = \delta_{i+1,j}$ we get $c_{j-1} = t_j$.

The spline $\sum_{i=-\infty}^{\infty} t_{i+1} B_i^1(x)$ is piecewise linear and agrees with the function x at all knots.

It is therefore equal to x.

6.3.20. In Equation (9) take all $c_i = 1$.

Then $d_i = 0$.

Hence, we obtain $d/dx \sum_{i=-\infty}^{\infty} B_i^k(x) = 0$ and $\sum_{i=-\infty}^{\infty} B_i^k(x)$ is constant.

6.3.24. Use Equation (14) with all A's zero except $A_j = 1$.

The result is $B_{j-2}^2(t_j) = (t_{j+1} - t_j)/(t_{j+1} - t_{j-1})$.

Next, take all A's zero except $A_{j+1} = 1$.

The result is $B_{j-1}^2(t_j) = (t_j - t_{j-1})/(t_{j+1} - t_{j-1})$.

6.3.28. The answer is No.

Consider four points $\tau_1 < \cdots < \tau_4$.

If $t_i < t_{i+1}$ and $\tau_i = \frac{1}{2}(t_{i+1} + t_{i+2})$, then these are necessary conditions:

$$\tau_1 < t_3,\ t_4 = 2\tau_2 - t_3 < 2\tau_2 - \tau_1,\ \tau_4 > t_5 = 2\tau_3 - t_4 > 2\tau_3 - 2\tau_2 + \tau_1$$

The τ-points 0, 1, 4, 5 violate the final condition.

6.3.30. Use Equation (7) to evaluate the spline.

Put $C_i^2 = t_{i+1}t_{i+2}$.

Then calculations produce

$$\begin{aligned} C_i^1 &= [C_i^2(x - t_i) + C_{i-1}^2(t_{i+2} - x)]/(t_{i+2} - t_i) = xt_{i+1} \\ C_i^0 &= [C_i^1(x - t_i) + C_{i-1}^1(t_{i+1} - x)]/(t_{i+1} - t_i) = x^2 \end{aligned}$$

6.3.32. It was shown in Section 6.3 that $B_i^k(x) = 0$ if and only if $x \notin (t_i, t_{i+k+1})$.

Hence, we obtain $B_i^k(t_j) = 0$ if and only if $t_j \geq t_{i+k+1}$ or $t_j \leq t_i$.

These conditions can be written $j \geq i + k + 1$ and $j \leq i$, respectively.

6.3.33. By Equation (8),

$$\frac{d}{dx}B_i^2(x) = \frac{2}{t_{i+2} - t_i}B_i^1(x) - \frac{2}{t_{i+3} - t_{i+1}}B_{i+1}^1(x)$$

This expression is zero at the points where

$$B_i^1(x)(t_{i+3} - t_{i+1}) = B_{i+1}^1(x)(t_{i+2} - t_i)$$

The nontrivial root is in the interval (t_{i+1}, t_{i+2}) and is obtained by solving

$$\frac{t_{i+2} - x}{t_{i+2} - t_{i+1}}(t_{i+3} - t_{i+1}) = \frac{x - t_{i+1}}{t_{i+2} - t_{i+1}}(t_{i+1} - t_{i+2})$$

The result is $x = (t_{i+3}t_{i+2} - t_it_{i+1})/(t_{i+3} + t_{i+2} - t_{i+1} - t_i)$

6.3 Computer Exercises

6.3.C1. $47,040$

Chapter 7

Initial-Value Problems

7.1 Taylor Series Method

7.1.1. **a.** $x = \frac{1}{4}t^4 + \frac{7}{3}t^3 - \frac{2}{3}t^{3/2} + c$

b. $x = ce^t$

e. $x = c_1 e^t + c_2 e^{-t}$ or $x = c_1 \cosh t + c_2 \sinh t$

7.1.2. **a.** $x = \frac{1}{3}t^3 + \frac{3}{4}t^{4/3} + 7$

7.1.3. **c.** $x = \int_0^t (\sin r/r)dr + c$

$$= \sum_{n=0}^{\infty}(-1)^n[(t^{2n+1})/((2n+1)(2n+1)!)] + c$$

d. $x = e^{(-t^2/2)}\Big[\int t^2 e^{(t^2/2)}dt + c\Big]$

7.1.4. ODE: $x' + tx = t^2$ (No initial value given, so we want the general solution.)

Try $x = a_0 + a_1 t + a_2 t^2 + a_3 t^3 + \cdots$.

Now we have $tx = a_0 t + a_1 t^2 + a_2 t^3 + a_3 t^4 + \cdots$ and $x' = a_1 + 2a_2 t + 3a_3 t^2 + 4a_4 t^3 + \cdots$.

The LHS of ODE becomes $x' + tx = a_1 + (a_0 + 2a_2)t + (a_1 + 3a_3)t^2 + (a_2 + 4a_4)t^3 + \cdots$.

We want $x' + tx = t^2$ so we identify coefficients of t^n as follows. $n = 0 : a_1 = 0$; $n = 1 : 2a_2 + a_0 = 0$; $n = 2 : 3a_3 + a_1 = 1$; $n = 3 : 4a_4 + a_2 = 0$; $n = 4 : 5a_5 + a_3 = 0$; $n = 5 : 6a_6 + a_4 = 0$; $n = 6 : 7a_7 + a_5 = 0$; etc.

General, a_0 arbitrary, $a_1 = 0$, $a_2 = -a_0/2$, $a_3 = 1/3$ and $n \geq 4 : a_n = -a_{n-2}/n$.

This implies that a_0 arbitrary, $a_2 = -a_0/2$, $a_4 = a_0/(4 \cdot 2)$, $a_6 = -a_0/(6 \cdot 4 \cdot 2)$, etc. and $a_1 = 0$, $a_3 = 1/3$, $a_5 = -1/(5 \cdot 3)$, $a_7 = 1/(7 \cdot 5 \cdot 3)$, etc.

Hence, the simplest solution is

$$\begin{aligned} x &= a_0 - \frac{a_0}{2}t^2 + \frac{a_0}{4 \cdot 2}t^4 - \frac{a_0}{6 \cdot 4 \cdot 2}t^6 + \cdots + \frac{t^3}{3} - \frac{t^5}{3 \cdot 5} + \frac{t^7}{3 \cdot 5 \cdot 7} - \cdots \\ &= a_0 + a_0 \sum_{n=1}^{\infty} (-1)^n \left(\frac{(2n-1)!}{2^{n-1}(2n)!} \right) t^{2n} + \sum_{n=1}^{\infty} (-1)^{n-1} \left(\frac{n! 2^n}{(2n+1)!} \right) t^{2n+1} \end{aligned}$$

7.1.6. Let $p(t) = a_0 + a_1 t + a_2 t^2 + \cdots$ so $p'(t) = a_1 + 2a_2 t + 3a_3 t^2 + \cdots$.

So $p(t) - p'(t) = (a_0 - a_1) + (a_1 - 2a_2)t + (a_2 - 3a_3)t^2 + \cdots$.

We want $p - p' = -2t + t^2 + t^3$.

Setting $a_0 - a_1 = 0$, $a_1 - 2a_2 = -2$, $a_2 - 3a_3 = 1$, $a_3 - 4a_4 = 1$, and $a_k - (k+1)a_{k+1} = 0$ for $k \geq 4$, we have $a_1 = a_0$, $a_2 = 1 + a_0/2$, $a_3 = a_0/6$, and $a_4 = (a_0/6 - 1)/4$, $a_5 = (a_0/6 - 1)/20$, $a_6 = (a_0/6 - 1)/120$, $\cdots$.

Letting $a_0 = 6$, we have $p(t) = 6 + 6t + 4t^2 + t^3$.

7.1.9. The initial-value problem

$$x' = x, \qquad x(0) = c$$

has solution $x(t) = ce^t$.

The initial-value problem

$$x' = x, \qquad x(0) = c + \varepsilon$$

has solution $x(t) = (c + \varepsilon)e^t$.

The difference (error) is εe^t.

If $t = 10$ this is $\varepsilon e^{10} = 2.2 \times 10^4 \varepsilon$.

If $t = 20$, this is $\varepsilon e^{20} = 4.8 \times 10^8 \varepsilon$ (overwhelming).

For $x' = -x$, effect of ε is negligible.

7.1.10.
$$\begin{aligned}
x' &= t^2 + x^3\\
x'' &= 2t + 3x^2x'\\
x''' &= 2 + 6x(x')^2 + 3x^2x''\\
x^{(iv)} &= 18xx'x'' + 6(x')^3 + 3x^2x'''\\
x(t+h) &= x + hx' + (h^2/2)x'' + (h^3/3!)x''' + (h^4/4!)x^{(iv)}.
\end{aligned}$$

7.1.11. **a.**
$$\begin{aligned}
x' &= x + e^x\\
x'' &= (1+e^x)x'\\
x''' &= (1+e^x)x'' + e^x(x')^2\\
x^{(iv)} &= (1+e^x)x''' + 3e^x x'x'' + e^x(x')^3
\end{aligned}$$

7.1.12.
$$\begin{aligned}
x'' &= x^2e^t + x', & x''(0) &= e^0 + 2 = 3\\
x''' &= 2xx'e^t + x^2e^t + x'', & x'''(0) &= 2(2)e^0 + e^0 + 3 = 8
\end{aligned}$$

$$t = 0, \quad x = 1, \quad x' = 2, \quad h = 0.1$$

$$\begin{aligned}
x(t+h) &= x + hx' + (h^2/2)x'' + (h^3/6)x'''\\
x(0.1) &= x(0) + hx'(0) + (h^2/2)x''(0) + (h^3/6)x'''(0)\\
&= 1 + (0.1)(2) + [(0.1)^2/2](3) + [(0.1)^3/6](8)\\
&= 1 + .2 + .015 + .00133 = 1.21633
\end{aligned}$$

7.1.14.

$n \leftarrow 20$
$s \leftarrow x^{(n)}$
for $i = 1$ **to** $n - 1$
  $s \leftarrow x^{(n-i)} + [h/\text{real}(n+1-i)]s$
end for
$s \leftarrow x + h[s]$

7.1 Computer Exercises

7.1.C1. $x(2.77) = 385.79118$

7.1.C2. **b.** $x(1.75) = 0.63299\,99833$

c. $x(5) = -0.20873\,51554$

7.1.C3. $x(10) = 22026.47$

7.1.C4. **a.** Error at $t = 1$ is 1.8×10^{-10}.

7.1.C5. $x(0) = 0.03245\,34427$

7.1.C7. Given $x' = tx$ with $x(0) = 1$.

Differentiating, we obtain $x'' = tx' + x$, $x''' = tx'' + 2x'$, $x^{(iv)} = tx''' + 3x''$, $x^{(v)} = tx^{(4)} + 4x'''$, and so on.

Thus, the recursive formula is $x^{(n)} = tx^{(n-1)} + (n-1)x^{(n-2)}$.

The numerical solution is $x(1) = 1.64872\,12691$.

7.1.C9. $x(0) = 1.67984\,09205 \times 10^{-3}$

7.1.C10. $x(0) = -3.75940\,73450$

7.2 Runge-Kutta Methods

7.2.2. **c.** $f(t,x) = +\sqrt{x/(1-t^2)}$

7.2.3. TS: $x' = -tx^2$, $x'' = -x^2 - 2txx'$.

At $t = 0$, $x = 2$: $x' = 0$, $x'' = -4$.

$x(t-h) = x - hx' + (h^2/2)x''$.

$x(-0.2) = 2 - 0.08 = 1.92$.

RK: $f(t,x) = -tx^2; t = 0, x = 2, h = -0.2$

$K_1 = hf(t,x) = 0$,

$K_2 = hf(t+h, x+K_1) = -0.2f(-0.2, 2) = -0.2[-(-0.2)(2)^2] = -0.16$

$x(-0.2) = x(0) + \frac{1}{2}(K_1 + K_2) = 2 + \frac{1}{2}(-0.16) = 1.92$.

7.2.5. **a.**

```
real function f(t, x)
real t, x
   f ← t²/(1 − t + 2x)
end function f
```

7.2.8. Solve $(df/dx) = e^{-x^2}$, $f(0) = 0$

```
program Main
interface external function f
real t ← 0; x ← 0; h ← 0.01
integer m ← 1001
call RK4(f, t, λ, h, m)
end program Main

real function f(t, x)
real t, x
  f ← e^{-x^2}
end function f
```

7.2.10. By Taylor series and Exercise 7.2.14, we have

$$\text{(A)} \quad x(t+h) = x + hx' + (h^2/2)x'' + (h^3/6)x''' + \cdots$$
$$= x + hf + (h^2/2)Df + (h^3/6)(D^2 f + f_x Df) + \cdots$$

By the two variable Taylor series, we have

$$f(t+\alpha h, x+\beta h f) = f + \alpha h f_t + \beta h f f_x + (1/2)(\alpha h)^2 f_{tt} + \alpha\beta h^2 f_{tx} + (1/2)(\beta h f)^2 f_{xx} + \cdots$$

and (5) gives

$$\text{(B)} \quad x(t+h) = x + w_1 hf + w_2 hf + w_2 \alpha h^2 f_t + w_2 \beta h^2 f f_x$$
$$+ (1/2) w_2 \alpha^2 h^3 f_{tt} + w_2 \alpha\beta h^3 f_{tx} + (1/2) w_2 \beta^2 h^3 f^2 f_{xx} + \cdots$$
$$= x + (w_1 + w_2) hf + w_2 h^2 [\alpha(\partial/\partial t) + \beta f(\partial/\partial x)] f$$
$$+ (w_2 h/2)[\alpha h(\partial/\partial t) + \beta h f(\partial/\partial x)]^2 f + \cdots$$

By requiring that $w_1 + w_2 = 1$, $\alpha = \beta$, $w_2\alpha = (1/2)$, then (A) and (B) agree through the first 3 terms.

The error term is

$$\frac{h^3}{6}(D^2 f + f_x Df) - \frac{h^3\alpha}{4}(D^2 f) = h^3\left(\frac{1}{6} - \frac{\alpha}{4}\right)D^2 f + \frac{h^3}{6} f_x Df$$

Here $x' = f$, $x'' = Df$, and $x''' = D^2 f + f_x Df$.

7.2.11. Total error $= 36n2^{-50} + 9nh^5 = (b-a)[36 \times 2^{-50} h^{-1} + 9h^4] = E$ since $h = (b-a)/n$. $(dE/dh) = 0$: $(b-a)[-36 \times 2^{-50} h^{-2} + 36h^3] = 0$ or $h^5 = 2^{-50}$.

Thus, we have $h = 2^{-10} = 1/1024$.

7.2.12. Let's make local truncation error $\leq 10^{-13}$.

Thus, we have $100h^5 \leq 10^{-13}$ or $h \leq 10^{-3}$.

So take $h = 10^{-3}$ and hope that the three extra digits will be enough to preserve 10-digit precision.

7.2.14. **b.** $x^{(iv)} = (d/dt)(D^2f) + (d/dt)(Df)f_x + Df_xDf$

But $x'' = Df$ so $(d/dt)(Df) = x''' = D^2f + f_xDf$ and

$$\begin{aligned}\frac{d}{dt}(D^2f) &= f_{ttt} + f_{ttx}f + 2\Big(\frac{df}{dt}\Big)f_{xt} + 2f\Big(\frac{df}{dt}xt\Big) + \Big(\frac{d}{dt}f^2\Big)f_{xx} + f^2\Big(\frac{d}{dt}f_{xx}\Big)\\ &= f_{ttt} + f_{ttx}f + 2f_{xt}(f_t + f_xf) + 2f(f_{xxt} + f_{xtx}f)\\ &\quad + f_{xx}(2f)(f_t + f_xf) + f^2(f_{xxt} + f_{xxx}f)\\ &= f_{ttt} + 3ff_{ttx} + 3f^2f_{txx} + f^3f_{xxx} + 2f_{tx}Df + 2ff_{xx}Df\\ &= D^3f + 2f_{tx}Df + 2ff_{xx}Df = D^3f + 2Df_xDf\end{aligned}$$

where $D^3 = (\partial^3/\partial t^3) + 3f(\partial^3/\partial x\partial t^2) + 3f^2(\partial^3/\partial t\partial x^2) + f^3(\partial^3/\partial x^2)$.

Hence, we obtain

$$\begin{aligned}x^{(iv)} &= D^3f + 2Df_xDf + f_xD^2f + f_x^2Df + Df_xDf\\ &= D^3f + f_xD^2f + 3Df_xDf + f_x^2Df\end{aligned}$$

7.2.15. $\varphi'(t) = f_1(x+th, y+tk)h + f_2(x+th, y+tk)k$

$\varphi''(t) = f_{11}(x+th, y+tk)h^2 + f_{12}(x+th, y+tk)hk$

$+f_{21}(x+th, y+tk)kh + f_{22}(x+th, y+tk)k^2$, and so on.

Hence, we obtain

$\varphi'(0) = f_1(x,y)h + f_2(x,y)k$

$\varphi'' = f_{11}(x,y)h^2 + 2f_{12}(x,y)hk + f_{22}(x,y)k^2$ and so on.

Thus, we have $f(x+th, y+tk) = \varphi(t) = \varphi(0) + t\varphi'(0) + (t^2/2)\varphi''(0) + \cdots$

$= f(x,y) + t[f_1(x,y)h + f_2(x,y)k] + (1/2)t^2[f_{11}(x,y)h^2 + 2f_{12}(x,y)hk + f_{22}(x,y)k^2] + \cdots$.

Now let $t = 1$ to get the usual form of Taylor's series in 2 variables.

7.2.17.

$f = \sin x + \cos y,\ \ @(0,0)$	$f = 1$
$f_x = \cos x,\ f_y = -\sin y,$	$f_x = 1,\ f_y = 0$
$f_{xy} = 0,$	$f_{xy} = 0$
$f_{xx} = -\sin x,\ f_{yy} = -\cos y,$	$f_{xx} = 0,\ f_{yy} = -1$
$f_{xxy} = f_{yyx} = 0,$	$f_{xxy} = f_{yyx} = f_{xyy} = 0$
$f_{xxx} = -\cos x,\ f_{yyy} = \sin y,$	$f_{xxx} = -1,\ f_{yyy} = 0$

By the previous exercise, the Taylor series about $(0,0)$ is

$$f(x,y) = f + (xf_x + yf_y) + (1/2!)(x^2f_{xx} + 2xyf_{xy} + h^2f_{yy})$$

$$+ (1/3!)(x^3f_{xxx} + 3x^2yf_{xxy} + 3xy^2f_{xyy} + y^3f_{yyy}) + \cdots$$

where the functions on the RHS are evaluated at $(0,0)$.

Now $\sin x + \cos x = 1 + x - (y^2/2!) - (x^3/3!) + \cdots$.

Recall $\sin x = x - (x^3/3!) + \cdots$ and $\cos y = 1 - (y^2/2!) + \cdots$.

Thus, the Taylor series of $f(x,y) = g(x) + h(y)$ about (a,b) is equal to the Taylor series of $g(x)$ about a plus that of $h(y)$ about b.

7.2.18. $f = y^2 - 3\ln x,$

$f_x = -3/x = -3x^{-1}, \qquad f_y = 2y$

$f_{xx} = 3x^{-2}, \qquad f_{yy} = 2$

$f_{xy} = 0$

$f(x+h, y+k) \approx f + (hf_x + kf_y) + (1/2)(h^2 f_{xx} + 2hkf_{xy} + k^2 f_{yy})$

$x = 1,\ y = 0$

$$\begin{aligned} f(1+h,k) &\approx 0 + h(-3) + k(0) + (1/2)(h^2 3 + 0 + k^2 2) \\ &= -3h + (3/2)h^2 + k^2 \end{aligned}$$

7.2.19. $f = e^{1-xy},$

$f_x = e^{1-xy}(-y), \qquad f_y = e^{1-xy}(-x)$

We have $f(x,y) = f(1,1) + (x-1)f_x(1,1) + (y-1)f_y(1,1) + \cdots$ and $e^{1-xy} \approx 1 - (x-1) - (y-1) = 3 - x - y.$

7.2.20. $f = xe^y,$

$f_x = e^y, \qquad f_y = xe^y$

$f_{xx} = 0, \qquad f_{yy} = xe^y$

$f_{xy} = e^y,$

$$\begin{aligned} f(x+h, y+k) &= f + hf_x + kf_y + (1/2)(h^2 f_{xx} + 2hkf_{xy} + k^2 f_{yy}) \\ &= xe^y + he^y + kxe^y + (1/2)h^2 \cdot 0 + hke^y + (1/2)k^2 xe^y \\ &= e^y[x(1 + k + (1/2)k^2) + h + hk] \end{aligned}$$

We have $A = 1 + k + (1/2)k^2$ and $B = h(1+k)$.

7.2.21. $f = (y-x)^{-1}, \qquad f_y = -(y-x)^{-2} = -f^2$

$f_x = (y-x)^{-2} = f^2, \qquad f_{yy} = 2(y-x)^{-3} = 2f^3$

$f_{xx} = 2(y-x)^{-3} = 2f^3, \qquad f_{xy} = -2(y-x)^{-3} = -2f^3$

$$\begin{aligned} f(x+h, y+k) &= f + (hf_x + kf_y) + (1/2)(h^2 f_{xx} + 2hkf_{xy} + k^2 f_{yy}) + \cdots \\ &= f + (h-k)f + (h-k)^2 f^3 + \cdots \end{aligned}$$

Hence, we obtain, $A = 1$, $B = h - k$, $C = (h-k)^2$.

7.2.22. $f = e^{x^2+y},$

$f_x = 2xe^{x^2+y} = 2xf, \qquad f_y = e^{x^2+y} = f$

$f_{xx} = 2e^{x^2+y} + 4x^2 e^{x^2+y} = 2(1+2x^2)f$

$f_{yy} = e^{x^2+y} = f, \qquad f_{xy} = 2xe^{x^2+y} = 2xf$

Now, we have

$$\begin{aligned} f(x+h, y+k) &= f + hf_x + kf_y + (1/2)(h^2 f_{xx} + 2hkf_{xy} + k^2 f_{yy}) + \cdots \\ &= f + 2xhf + kf + (1+2x^2)h^2 f + 2hkxf + (1/2)k^2 f + \cdots \\ &\approx (1 + 2xh + k + (1+2x^2)h^2 + 2hkx + (1/2)k^2)f \end{aligned}$$

Thus, we have $f(h, 1-k) \approx (1 - k + h^2 + (1/2)k^2)f(0,1)$ and $f(.001, .998) \approx (.998003)e = 2.7128534.$

7.2 Computer Exercises

7.2.C2. $x(1) = 1.5708$ (true soln.); 1.519125 (computed soln.)

7.2.C3. **b.** $n = 7$, $x(2) = 0.82356\ 78972$ (RK), $x(2) = 0.82356\ 78970$ (TS)

c. $n = 7$, $x(2) = -0.49999\ 99998$ (RK), $x(2) = -0.50000\ 00012$ (TS)

7.2.C4. $x(1) = 0.60653 = x(3)$

7.2.C5. $x(3) = 1.5$

7.2.C6. $x(0.0) = 1.0 = x(1.6)$

7.2.C8. $x(1) = 3.95249$

7.2.C9. $x(10) = 1.344 \times 10^{43}$

7.3 Adaptive Runge-Kutta and Multistep Methods

7.3.1. $x_{k+1} = x_k + \frac{h}{2}[f(t, x(t)) + f(t+h, x(t+h))$
$= x_k + \frac{h}{2}[-x(t) - x(t+h)]$
$= x_k + \frac{h}{2}[-x_k - x_{k+1}]$

So $x_{k+1}(1 + h/2) = x_k(1 - h/2)$ and $x_{k+1} = cx_k$ where $c = (1 - h/2)/(1 + h/2)$

Thus, we have $x_0 = 1$, $x_1 = c$, $x_2 = c^2$, ... Let $h = 1/n$.

Then $x(1) = e^{-1}$ (true solution) and $x_n = \{[1 - 1/(2n)]/[1 + 1/(2n)]\}^n$ approximate solution.

Write $x_n = [(1-\epsilon)/(1+\epsilon)]^n = [(1-\epsilon)\sum_{k=0}^{\infty}(-\epsilon)^k]^n = [(1-\epsilon)(1-\epsilon+\epsilon^2-\epsilon^3+\cdots)]^n$ $= (1 - 2\epsilon + 2\epsilon^2 + \cdots)^n \approx (1-2\epsilon)^n = (1-1/n)^n \to \epsilon^{-1}$.

Error is $x(1) - x_n = e^{-1} - \{[1 - 1/(2n)]/[1 + 1/(2n)]\}^n = e^{-1} - [(2n-1)/(2n+1)]^n$.

For $n = 5$, for example, the error is $e^{-1} - (0.9/11)^5 = 0.36787\,94412 - 0.36664\,78321$ $= 0.00123\,1$.

7.3.2. $\int_{t-h}^{t+h} dx = \int_{t-h}^{t+h} f(r, x(r))\, dr$.

Using Simpson's rule, Equation (1) in Section 5.4, we have

$x(t+h) = x(t-h) + \dfrac{h}{3}[f(t-h, x(t-h)) + 4f(t, x(t)) + f(t+h, x(t+h))]$

7.3.4. Let LHS $= x(h)$ and RHS $= ax(0) + bx(-h) + h[cx'(h) + dx''(0) + ex'''(-h)]$

x	x'	x''	x'''	LHS	RHS
1	0	0	0	$1 =$	$a + b$
t	1	0	0	$h =$	$-hb + hc$
t^2	$2t$	2	0	$h^2 =$	$h^2b + 2h^2c + 2hd$
t^3	$3t^2$	$6t$	6	$h^3 =$	$-h^3b + 3h^3c + 6he$
t^4	$4t^3$	$12t^2$	$24t$	$h^4 =$	$h^4b + 4h^4c - 24h^2e$

Setting LHS = RHS.

Eliminating the e-term from the last two equations, we have $5h^4 = -3h^4b + 16h^4c$.

Now solving for b and c with this equation and the second one, we have $c = (2/13)$, $b = -(11/13)$. It follows that $a = (24/13)$, $d = (10/13)h$, $e = -(2/39)h^2$.

7.3.5. Let LHS $= x(h)$ and RHS $= ax(0) + bx'(0) + cx'(h)$

x	x'	LHS	RHS
1	0	$1 =$	a
t	1	$h =$	$b + c$
t^2	$2t$	$h^2 =$	$2hc$

Letting LHS = RHS, we have $a = 1$, $b = c = h/2$.

Now LHS $= x + hx' + (h^2/2)x'' + (h^3/3!)x''' + \cdots$ by Taylor series and

$$\begin{aligned}\text{RHS} &= x + (h/2)x' + (h/2)(x' + hx'' + (h^2/2)x'' + \cdots) \\ &= x + hx' + (h^2/2)x'' + (h^3/4)x''' + \cdots\end{aligned}$$

The error term is $\mathcal{O}(h^3)$ since LHS $\neq$ RHS after the third term.

7.3.8. Consider $x' = (1 + t^2)x$, $x(0) = s$.

Since $(x'/x) = 1 + t^2$, we have $\ln x = t + (t^3/3) + c$ and the solution is $x = se^{t+t^3/3}$ since $x(0) = s$.

Now $x(9) = se^{252}$ and $(\partial/\partial s)x(9, s) = e^{252} \approx 10^{109}$.

Thus, it is very sensitive.

7.3.9. **a.** $f = \sin t + e^x$; $f_x = e^x > 0$; divergent for all t.

c. $f = xt$; $f_x = t > 0$ if $t > 0$; divergent for positive t.

e. $f = \cos t - e^x$; $f_x = -e^x < 0$; divergent for no t.

7.3.11. $f = (1 + t^2)^{-1}x$; $f_x = (1 + t^2)^{-1} > 0$; divergent for all t.

7.3 Computer Exercises

7.3.C5. $x(1/2) = 2.25$

7.3.C6. $x(-1/2) = -4.5$

7.3.C9. $y(e) = -6.38905\ 60989$ where $y(x) = [1 - \ln v(x)]v(x)$

7.3.C12. 0.21938 39244

7.3.C13. 0.99530 87432

7.3.C15. Si(1) = 0.94608 30703

7.4 Methods for First and Higher-Order Systems

7.4.1. We determine the derivatives

$$\begin{aligned} x' &= y, & y' &= x \\ x'' &= y' = x, & y'' &= x' = y \\ x''' &= x' = y, & y''' &= y' = x \\ x^{(iv)} &= y' = x, & y^{(iv)} &= x' = y \\ x^{(v)} &= x' = y, & y^{(v)} &= y' = x \end{aligned}$$

The Taylor series formulas are

$$\begin{aligned} x(t+h) &= x + hx' + \tfrac{1}{2}h^2x'' + \tfrac{1}{3!}h^3x''' + \tfrac{1}{4!}h^4x^{(iv)} + \tfrac{1}{5!}h^5x^{(v)} \\ &= x(1 + \tfrac{1}{2}h^2 + \tfrac{1}{24}h^4) + y(h + \tfrac{1}{6}h^3 + \tfrac{1}{120}h^5) \\ y(t+h) &= y + hy' + \tfrac{1}{2}h^2y'' + \tfrac{1}{3!}h^3y''' + \tfrac{1}{4!}h^4y^{(iv)} + \tfrac{1}{5!}h^5y^{(v)} \\ &= y(1 + \tfrac{1}{2}h^2 + \tfrac{1}{24}h^4) + x(h + \tfrac{1}{6}h^3 + \tfrac{1}{120}h^5) \end{aligned}$$

7.4.2. Since system is not coupled, solve two separate problems.

7.4.3. System is not coupled so each differential equation can be solved individually.

7.4.4. Let $x_0 = t, x_1 = x, x_2 = y$

$$\mathbf{X}' = \begin{bmatrix} 1 \\ x_1^2 + \log x_2 + x_0^2 \\ e^{x_2} - \cos x_1 + \sin(x_0x_1) - (x_1x_2)^7 \end{bmatrix}, \qquad \mathbf{X}(0) = [0, 1, 3]^T$$

7.4.5. Let $x_1 = x$, $x_2 = x'$, $x_3 = x''$

$$\mathbf{X}' = \begin{bmatrix} x_2 \\ x_3 \\ 2x_2 + \log x_3 + \cos x_1 \end{bmatrix}, \qquad \mathbf{X}(0) = [1, -3, 5]^T$$

Or in autonomous form: Let $x_0 = t$, $x_1 = x$, $x_2 = x'$, $x_3 = x''$

$$\mathbf{X}' = \begin{bmatrix} 1 \\ x_2 \\ x_3 \\ 2x_2 + \log x_3 + \cos x_1 \end{bmatrix}, \qquad \mathbf{X}(0) = [0, 1, -3, 5]^T$$

7.4.7. Solve each equation separately since they are not coupled.

7.4.8. Let $x_1 = x$, $x_2 = x'$, $x_3 = y$, $x_4 = y'$. Then

$$\mathbf{X}' = \begin{bmatrix} x_2 \\ -x_1(x_1^2 + x_3^2)^{-3/2} \\ x_4 \\ -x_3(x_1^2 + x_3^2)^{-3/2} \end{bmatrix}, \qquad \mathbf{X}(0) = \begin{bmatrix} 0.5 \\ 0.75 \\ 0.25 \\ 1.0 \end{bmatrix}$$

Or in autonomous form: Let $x_0 = t$, $x_1 = x$, $x_2 = x'$, $x_3 = y$, $x_4 = y'$. Then

$$\mathbf{X}' = \begin{bmatrix} 1 \\ x_2 \\ -x_1(x_1^2 + x_3^2)^{-3/2} \\ x_4 \\ -x_3(x_1^2 + x_3^2)^{-3/2} \end{bmatrix}, \qquad \mathbf{X}(0) = \begin{bmatrix} 0 \\ 0.5 \\ 0.75 \\ 0.25 \\ 1.0 \end{bmatrix}$$

7.4.9. Let $x_1 = x$, $x_2 = x'$, $x_3 = x''$, $x_4 = x'''$

$$\mathbf{X}' = \begin{bmatrix} x_2 \\ x_3 \\ x_4 \\ x_4^2 + \cos(x_2 x_3) - \sin(t x_1) + \log(x_1/t) \end{bmatrix}, \qquad \mathbf{X}(0) = [1, 3, 4, 5]^T$$

Or in autonomous form: Let $x_0 = t$, $x_1 = x$, $x_2 = x'$, $x_3 = x''$, $x_4 = x'''$

$$\mathbf{X}' = \begin{bmatrix} 1 \\ x_2 \\ x_3 \\ x_4 \\ x_4^2 + \cos(x_2 x_3) - \sin(x_0 x_1) + \log(x_1/x_0) \end{bmatrix}, \qquad \mathbf{X}(0) = [0, 1, 3, 4, 5]^T$$

7.4.10. Let $x_1 = x$, $x_2 = y$, $x_3 = z$, $x_4 = x'$, $x_5 = y'$, $x_6 = z'$. Then

$$\mathbf{X}' = \begin{bmatrix} x_4 \\ x_5 \\ x_6 \\ 2x_1x_3x_4 + 3x_1^2x_2t^2 \\ e^{x_2}x_5 + 4x_1t^2x_3 \\ 2tx_6 + 2te^{x_1x_3} \end{bmatrix}, \qquad \mathbf{X}(1) = \begin{bmatrix} 3 \\ 3 \\ 2 \\ -79/12 \\ 2 \\ 3 \end{bmatrix}$$

Since $2 - 2(3)(2)x' = 3(9)(3)(1)$.

Or in autonomous form: Let $x_0 = t$, $x_1 = x$, $x_2 = y$, $x_3 = z$, $x_4 = x'$, $x_5 = y'$, $x_6 = z'$. Then

$$\mathbf{X}' = \begin{bmatrix} 1 \\ x_4 \\ x_5 \\ x_6 \\ 2x_1x_3x_4 + 3x_1^2x_2x_0^2 \\ e^{x_2}x_5 + 4x_1x_0^2x_3 \\ 2x_0x_6 + 2x_0e^{x_1x_3} \end{bmatrix}, \qquad \mathbf{X}(1) = \begin{bmatrix} 1 \\ 3 \\ 3 \\ 2 \\ -79/12 \\ 2 \\ 3 \end{bmatrix}$$

7.4.11. **a.** Let $x_1 = x$, $x_2 = x'$, $x_3 = x''$. Then

$$\mathbf{X}' = \begin{bmatrix} x_2 \\ x_3 \\ -x_3 \sin x_1 - tx_2 - x_3 \end{bmatrix}$$

Or in autonomous form: Let $x_0 = t$, $x_1 = x$, $x_2 = x'$, $x_3 = x''$. Then

$$\mathbf{X}' = \begin{bmatrix} 1 \\ x_2 \\ x_3 \\ -x_3 \sin x_1 - x_0x_2 - x_3 \end{bmatrix}$$

7.4.12. Let $x_1 = x$, $x_2 = x'$

$$\mathbf{X}' = \begin{bmatrix} x_2 \\ x_2 - x_1 \end{bmatrix}, \qquad \mathbf{X}(0) = [0, 1]^T$$

Or in autonomous form: Let $x_0 = t$, $x_1 = x$, $x_2 = x'$

$$\mathbf{X}' = \begin{bmatrix} 1 \\ x_2 \\ x_2 - x_1 \end{bmatrix}, \qquad \mathbf{X}(0) = [0, 0, 1]^T$$

7.4.13. Let $x_0 = t$, $x_1 = x$, $x_2 = y$, $x_3 = x'$, $x_4 = y'$

$$\mathbf{X}' = \begin{bmatrix} 1 \\ x_3 \\ x_4 \\ x_1 + x_2 - 2x_3 + 3x_4 + \log x_0 \\ 2x_1 - 3x_2 + 5x_3 + x_0 x_2 - \sin x_0 \end{bmatrix}, \qquad \mathbf{X}(0) = [0, 1, 3, 2, 4]^T$$

7.4 Computer Exercises

7.4.C1. $x(1) = 2.46869\ 39399$, $y(1) = 1.28735\ 52872$

7.4.C2. $x(0.38) = 1.90723 \times 10^{12}$, $y(0.38) = -8.28807 \times 10^4$

7.4.C4. $x(-1) = 3.36788$, $y(-1) = 2.36788$

7.4.C5. $x_1\left(\frac{\pi}{2}\right) = x_4\left(\frac{\pi}{2}\right) = 0, x_2\left(\frac{\pi}{2}\right) = 1, x_3\left(\frac{\pi}{2}\right) = -1$

7.4.C7. $x(6) = 4.39411$, $y(6) = 3.10378$

7.5 Adams-Bashforth-Moulton Methods

7.5.1. The new system is $x_1' = x_2, x_2' = \alpha x_3, x_3' = x_4, x_4' = \beta x_1$.

The characteristic equation for the 4×4 matrix is $\lambda^4 = \alpha\beta$.

Since the eigenvalues are distinct, the matrix exponential e^{At} is the diagonal matrix having elements $e^{\lambda_j t}$, and the solution functions are

$$x_j(t) = e^{\lambda_j t} x_j(0)$$

Chapter 8

More on Linear Systems

8.1 Matrix Factorizations

8.1.1. a.

$$\mathbf{L} = \begin{bmatrix} 1 & 0 & 0 \\ 0 & 1 & 0 \\ \frac{1}{3} & -3 & 1 \end{bmatrix}, \quad \mathbf{U} = \begin{bmatrix} 3 & 0 & 3 \\ 0 & -1 & 3 \\ 0 & 0 & 8 \end{bmatrix}$$

8.1.2. a.

$$\mathbf{A} = \begin{bmatrix} 1 & 0 & 0 & 0 \\ 0 & 1 & 0 & 0 \\ 0 & 3 & 1 & 0 \\ 5 & 0 & 2 & 1 \end{bmatrix} \begin{bmatrix} 1 & 0 & 0 & 2 \\ 0 & 3 & 0 & 0 \\ 0 & 0 & 4 & 0 \\ 0 & 0 & 0 & 0 \end{bmatrix} = \mathbf{LU}$$

$$\mathbf{ML} = \begin{bmatrix} 1 & 0 & 0 & 0 \\ 0 & 1 & 0 & 0 \\ 0 & -3 & 1 & 0 \\ -5 & 6 & -2 & 1 \end{bmatrix} \begin{bmatrix} 1 & 0 & 0 & 0 \\ 0 & 1 & 0 & 0 \\ 0 & 3 & 1 & 0 \\ 5 & 0 & 2 & 1 \end{bmatrix} = \begin{bmatrix} 1 & 0 & 0 & 0 \\ 0 & 1 & 0 & 0 \\ 0 & 0 & 1 & 0 \\ 0 & 0 & 0 & 1 \end{bmatrix}$$

8.1.3. **a.**

$$\mathbf{LU} = \begin{bmatrix} 1 & 0 & 0 & 0 & 0 \\ 0 & 1 & 0 & 0 & 0 \\ 0 & 2 & 1 & 0 & 0 \\ 0 & 4 & 2 & 1 & 0 \\ 4 & 0 & 0 & 0 & 1 \end{bmatrix} \begin{bmatrix} 25 & 0 & 0 & 0 & 1 \\ 0 & 27 & 4 & 3 & 2 \\ 0 & 0 & 50 & -6 & -4 \\ 0 & 0 & 0 & 0 & 0 \\ 0 & 0 & 0 & 0 & 20 \end{bmatrix}$$

$$\mathbf{M} = \mathbf{M}_4\mathbf{M}_3\mathbf{M}_2\mathbf{M}_1$$

$$= \begin{bmatrix} 1 & 0 & 0 & 0 & 0 \\ 0 & 1 & 0 & 0 & 0 \\ 0 & 0 & 1 & 0 & 0 \\ 0 & 0 & 0 & 1 & 0 \\ 0 & 0 & 0 & 0 & 1 \end{bmatrix} \begin{bmatrix} 1 & 0 & 0 & 0 & 0 \\ 0 & 1 & 0 & 0 & 0 \\ 0 & 0 & 1 & 0 & 0 \\ 0 & 0 & -2 & 1 & 0 \\ 0 & 0 & 0 & 0 & 1 \end{bmatrix} \begin{bmatrix} 1 & 0 & 0 & 0 & 0 \\ 0 & 1 & 0 & 0 & 0 \\ 0 & -2 & 1 & 0 & 0 \\ 0 & -4 & 0 & 1 & 0 \\ 0 & 0 & 0 & 0 & 1 \end{bmatrix} \begin{bmatrix} 1 & 0 & 0 & 0 & 0 \\ 0 & 1 & 0 & 0 & 0 \\ 0 & 0 & 1 & 0 & 0 \\ 0 & 0 & 0 & 1 & 0 \\ -4 & 0 & 0 & 0 & 0 \end{bmatrix}$$

$$= \begin{bmatrix} 1 & 0 & 0 & 0 & 0 \\ 0 & 1 & 0 & 0 & 0 \\ 0 & 0 & 1 & 0 & 0 \\ 0 & 0 & -2 & 1 & 0 \\ 0 & 0 & 0 & 0 & 1 \end{bmatrix} \begin{bmatrix} 1 & 0 & 0 & 0 & 0 \\ 0 & 1 & 0 & 0 & 0 \\ 0 & -2 & 1 & 0 & 0 \\ 0 & -4 & 0 & 1 & 0 \\ -4 & 0 & 0 & 0 & 1 \end{bmatrix} = \begin{bmatrix} 1 & 0 & 0 & 0 & 0 \\ 0 & 1 & 0 & 0 & 0 \\ 0 & -2 & 1 & 0 & 0 \\ 0 & 0 & -2 & 1 & 0 \\ -4 & 0 & 0 & 0 & 1 \end{bmatrix}$$

8.1.4. **b.** Let

$$\mathbf{A} = \begin{bmatrix} 3 & 2 & 1 \\ 2 & 2 & 1 \\ 1 & 1 & 1 \end{bmatrix} \Rightarrow \begin{bmatrix} 3 & 2 & 1 \\ 0 & \frac{2}{3} & \frac{1}{3} \\ 0 & \frac{1}{3} & \frac{2}{3} \end{bmatrix} \Rightarrow \begin{bmatrix} 3 & 2 & 1 \\ 0 & \frac{2}{3} & \frac{1}{3} \\ 0 & 0 & \frac{1}{2} \end{bmatrix}$$

Now

$$\mathbf{A} = \begin{bmatrix} 1 & 0 & 0 \\ \frac{2}{3} & 1 & 0 \\ \frac{1}{3} & \frac{1}{2} & 1 \end{bmatrix} \begin{bmatrix} 3 & 2 & 1 \\ 0 & \frac{2}{3} & \frac{1}{3} \\ 0 & 0 & \frac{1}{2} \end{bmatrix}$$

8.1.5. **a.**

$$\mathbf{M}_1 = \begin{bmatrix} 1 & 0 & 0 & 0 \\ 0 & 1 & 0 & 0 \\ 0 & 0 & 1 & 0 \\ -\frac{w}{a} & 0 & 0 & 1 \end{bmatrix}, \quad \mathbf{M}_2 = \begin{bmatrix} 1 & 0 & 0 & 0 \\ 0 & 1 & 0 & 0 \\ 0 & -\frac{x}{b} & 1 & 0 \\ 0 & 0 & 0 & 1 \end{bmatrix}, \quad \mathbf{M}_3 = \begin{bmatrix} 1 & 0 & 0 & 0 \\ 0 & 1 & 0 & 0 \\ 0 & 0 & 1 & 0 \\ 0 & 0 & -\frac{y}{c} & 1 \end{bmatrix}$$

$$\mathbf{M} = \mathbf{M}_3\mathbf{M}_2\mathbf{M}_1 = \begin{bmatrix} 1 & 0 & 0 & 0 \\ 0 & 1 & 0 & 0 \\ 0 & -\frac{x}{b} & 1 & 0 \\ -\frac{w}{a} & \frac{xy}{bc} & -\frac{y}{c} & 1 \end{bmatrix}, \quad \mathbf{U} = \begin{bmatrix} a & 0 & 0 & z \\ 0 & b & 0 & 0 \\ 0 & 0 & c & 0 \\ 0 & 0 & 0 & d - \frac{wz}{a} \end{bmatrix}$$

b.

$$\mathbf{LU} = \begin{bmatrix} 1 & 0 & 0 & 0 \\ 0 & 1 & 0 & 0 \\ 0 & \frac{x}{b} & 1 & 0 \\ \frac{w}{a} & 0 & \frac{y}{c} & 1 \end{bmatrix} \begin{bmatrix} a & 0 & 0 & z \\ 0 & b & 0 & 0 \\ 0 & 0 & c & 0 \\ 0 & 0 & 0 & d - \frac{wz}{a} \end{bmatrix}$$

$$\mathbf{A} = \mathbf{L}'\mathbf{U}' = \begin{bmatrix} a & 0 & 0 & 0 \\ 0 & b & 0 & 0 \\ 0 & x & c & 0 \\ 0 & 0 & y & d - \frac{wz}{a} \end{bmatrix} \begin{bmatrix} 1 & 0 & 0 & \frac{z}{a} \\ 0 & 1 & 0 & 0 \\ 0 & 0 & 1 & 0 \\ 0 & 0 & 0 & 1 \end{bmatrix}$$

8.1.6. **a.**

$$\mathbf{LU} = \begin{bmatrix} 1 & 0 & 0 & 0 \\ -\frac{1}{4} & 1 & 0 & 0 \\ -\frac{1}{4} & -\frac{1}{15} & 1 & 0 \\ 0 & -\frac{4}{15} & -\frac{2}{7} & 1 \end{bmatrix} \begin{bmatrix} 4 & -1 & -1 & 0 \\ 0 & \frac{15}{4} & -\frac{1}{4} & -1 \\ 0 & 0 & \frac{56}{15} & -\frac{16}{15} \\ 0 & 0 & 0 & \frac{24}{7} \end{bmatrix}$$

b.

$$\mathbf{U} = \mathbf{D}\mathbf{U}' = \begin{bmatrix} 4 & 0 & 0 & 0 \\ 0 & \frac{15}{4} & 0 & 0 \\ 0 & 0 & \frac{56}{15} & 0 \\ 0 & 0 & 0 & \frac{24}{7} \end{bmatrix} \begin{bmatrix} 1 & -\frac{1}{4} & -\frac{1}{4} & 0 \\ 0 & 1 & -\frac{1}{15} & -\frac{4}{15} \\ 0 & 0 & 1 & -\frac{2}{7} \\ 0 & 0 & 0 & 1 \end{bmatrix}$$

c.

$$\mathbf{L}' = \begin{bmatrix} 4 & 0 & 0 & 0 \\ -1 & \frac{15}{4} & 0 & 0 \\ -1 & -\frac{1}{4} & \frac{56}{15} & 0 \\ 0 & -1 & -\frac{16}{15} & \frac{24}{7} \end{bmatrix}$$

d.

$$\mathbf{L}'' = \mathbf{L}\begin{bmatrix} 2 & 0 & 0 & 0 \\ 0 & \frac{\sqrt{15}}{2} & 0 & 0 \\ 0 & 0 & 2\sqrt{\frac{14}{15}} & 0 \\ 0 & 0 & 0 & 2\sqrt{\frac{6}{7}} \end{bmatrix} = \begin{bmatrix} 2 & 0 & 0 & 0 \\ -\frac{1}{2} & \frac{1}{2}\sqrt{15} & 0 & 0 \\ -\frac{1}{2} & -1/(2\sqrt{15}) & 2\sqrt{\frac{14}{15}} & 0 \\ 0 & -2/(\sqrt{15}) & -\frac{4}{7}\sqrt{\frac{14}{15}} & 2\sqrt{\frac{6}{7}} \end{bmatrix}$$

e. $\mathrm{Det}(\mathbf{A}) = \mathrm{Det}(\mathbf{D}) = 4 \cdot \frac{15}{4} \cdot \frac{56}{15} \cdot \frac{24}{7} = \frac{1344}{7} = 192.$

8.1.8.

$$\mathbf{M}_1 = \begin{bmatrix} 1 & 0 & 0 & 0 \\ -1 & 1 & 0 & 0 \\ 1 & 0 & 1 & 0 \\ -1 & 0 & 0 & 1 \end{bmatrix}, \quad \mathbf{M}_2 = \begin{bmatrix} 1 & 0 & 0 & 0 \\ 0 & 1 & 0 & 0 \\ 0 & -1 & 1 & 0 \\ 0 & 1 & 0 & 1 \end{bmatrix}, \quad \mathbf{M}_3 = \begin{bmatrix} 1 & 0 & 0 & 0 \\ 0 & 1 & 0 & 0 \\ 0 & 0 & 1 & 0 \\ 0 & 0 & -1 & 1 \end{bmatrix}$$

$$\mathbf{M}_3\mathbf{M}_2\mathbf{M}_1\mathbf{A} = \begin{bmatrix} 1 & 0 & 0 & 1 \\ 0 & 1 & 0 & -2 \\ 0 & 0 & 1 & 4 \\ 0 & 0 & 0 & -8 \end{bmatrix} = \mathbf{U}$$

$$\mathbf{M}_1^{-1}\mathbf{M}_2^{-1}\mathbf{M}_3^{-1} = \begin{bmatrix} 1 & 0 & 0 & 0 \\ 1 & 1 & 0 & 0 \\ -1 & 0 & 1 & 0 \\ 1 & 0 & 0 & 1 \end{bmatrix}\begin{bmatrix} 1 & 0 & 0 & 0 \\ 0 & 1 & 0 & 0 \\ 0 & 1 & 1 & 0 \\ 0 & -1 & 0 & 1 \end{bmatrix}\begin{bmatrix} 1 & 0 & 0 & 0 \\ 0 & 1 & 0 & 0 \\ 0 & 0 & 1 & 0 \\ 0 & 0 & 1 & 1 \end{bmatrix}$$

$$= \begin{bmatrix} 1 & 0 & 0 & 0 \\ 1 & 1 & 0 & 0 \\ -1 & 1 & 1 & 0 \\ 1 & -1 & 1 & 1 \end{bmatrix} \equiv \mathbf{L}$$

8.1.9. a.

$$\mathbf{L} = \begin{bmatrix} 1 & 0 & 0 \\ 1 & 1 & 0 \\ 3 & -1 & 1 \end{bmatrix}, \quad \mathbf{D} = \begin{bmatrix} 2 & 0 & 0 \\ 0 & -2 & 0 \\ 0 & 0 & 3 \end{bmatrix}, \quad \mathbf{U}' = \begin{bmatrix} 1 & -\frac{1}{2} & 1 \\ 0 & 1 & -\frac{1}{2} \\ 0 & 0 & 1 \end{bmatrix}$$

b. $\mathbf{LDU'x} = \mathbf{b} \Rightarrow \mathbf{U'x} = \mathbf{D}^{-1}\mathbf{L}^{-1}b = \mathbf{D}^{-1}\mathbf{M}_2\mathbf{M}_1\mathbf{b}$

$$\mathbf{D}^{-1}\mathbf{M}_2\mathbf{M}_1\mathbf{b} = \begin{bmatrix} \frac{1}{2} & 0 & 0 \\ 0 & -\frac{1}{2} & 0 \\ 0 & 0 & \frac{1}{3} \end{bmatrix} \begin{bmatrix} 1 & 0 & 0 \\ 0 & 1 & 0 \\ 0 & 1 & 1 \end{bmatrix} \begin{bmatrix} 1 & 0 & 0 \\ -1 & 1 & 0 \\ -3 & 0 & 1 \end{bmatrix} \begin{bmatrix} -2 \\ -5 \\ 0 \end{bmatrix}$$

$$= \begin{bmatrix} \frac{1}{2} & 0 & 0 \\ 0 & -\frac{1}{2} & 0 \\ 0 & 0 & \frac{1}{3} \end{bmatrix} \begin{bmatrix} 1 & 0 & 0 \\ -1 & 1 & 0 \\ -4 & 1 & 1 \end{bmatrix} \begin{bmatrix} -2 \\ -5 \\ 0 \end{bmatrix} = \begin{bmatrix} -1 \\ \frac{3}{2} \\ 1 \end{bmatrix}$$

Hence, we obtain

$$\begin{bmatrix} 1 & -\frac{1}{2} & 1 \\ 0 & 1 & -\frac{1}{2} \\ 0 & 0 & 1 \end{bmatrix} \begin{bmatrix} x_1 \\ x_2 \\ x_3 \end{bmatrix} = \begin{bmatrix} -1 \\ \frac{3}{2} \\ 1 \end{bmatrix}$$

$$\Rightarrow x_3 = 1, \quad x_2 = 2, \quad x_1 = -1 \text{ or } \mathbf{x} = (-1, 2, 1)^T$$

8.1.10. **a.**

$$\mathbf{L} = \begin{bmatrix} 1 & 0 & 0 \\ 2 & 1 & 0 \\ -1 & 3 & 1 \end{bmatrix}, \quad \mathbf{D} = \begin{bmatrix} -2 & 0 & 0 \\ 0 & 1 & 0 \\ 0 & 0 & -1 \end{bmatrix}, \quad \mathbf{U'} = \begin{bmatrix} 1 & -\frac{1}{2} & 1 \\ 0 & 1 & 1 \\ 0 & 0 & 1 \end{bmatrix}$$

b.

$$\mathbf{D}^{-1}\mathbf{M}_2\mathbf{M}_1\mathbf{b} = \begin{bmatrix} -\frac{1}{2} & 0 & 0 \\ 0 & 1 & 0 \\ 0 & 0 & -1 \end{bmatrix} \begin{bmatrix} 1 & 0 & 0 \\ 0 & 1 & 0 \\ 0 & -3 & 1 \end{bmatrix} \begin{bmatrix} 1 & 0 & 0 \\ -2 & 1 & 0 \\ 1 & 0 & 1 \end{bmatrix} \begin{bmatrix} 1 \\ 4 \\ 4 \end{bmatrix}$$

$$= \begin{bmatrix} -\frac{1}{2} & 0 & 0 \\ 0 & 1 & 0 \\ 0 & 0 & -1 \end{bmatrix} \begin{bmatrix} 1 & 0 & 0 \\ -2 & 1 & 0 \\ 7 & -3 & 1 \end{bmatrix} \begin{bmatrix} 1 \\ 4 \\ 4 \end{bmatrix} = \begin{bmatrix} -\frac{1}{2} & 0 & 0 \\ -2 & 1 & 0 \\ -7 & 3 & -1 \end{bmatrix} \begin{bmatrix} 1 \\ 4 \\ 4 \end{bmatrix} = \begin{bmatrix} -\frac{1}{2} \\ 2 \\ 1 \end{bmatrix}$$

Hence, we obtain

$$\begin{bmatrix} 1 & -\frac{1}{2} & 1 \\ 0 & 1 & 1 \\ 0 & 0 & 1 \end{bmatrix} \begin{bmatrix} x_1 \\ x_2 \\ x_3 \end{bmatrix} = \begin{bmatrix} -\frac{1}{2} \\ 2 \\ 1 \end{bmatrix}$$

$$\Rightarrow x_3 = 1, \quad x_2 = 1, \quad x_1 = -1, \text{ or } \mathbf{x} = [-1, 1, 1]^T$$

8.1.12. Solve:

$$\begin{bmatrix} 3 & 2 & -1 \\ 0 & -\frac{1}{3} & \frac{11}{3} \\ 0 & 0 & 15 \end{bmatrix} \begin{bmatrix} x_1^{(1)} \\ x_2^{(1)} \\ x_3^{(1)} \end{bmatrix} = \begin{bmatrix} 1 & 0 & 0 \\ -\frac{5}{3} & 1 & 0 \\ -8 & 5 & 1 \end{bmatrix} \begin{bmatrix} 1 \\ 0 \\ 0 \end{bmatrix} = \begin{bmatrix} 1 \\ -\frac{5}{3} \\ -8 \end{bmatrix}$$

Hence, we have $x_3^{(1)} = -\frac{8}{15}, \quad x_2^{(1)} = -\frac{13}{15}, \quad x_1^{(1)} = \frac{11}{15}$

Solve:

$$\begin{bmatrix} 3 & 2 & -1 \\ 0 & -\frac{1}{3} & \frac{11}{3} \\ 0 & 0 & 15 \end{bmatrix} \begin{bmatrix} x_1^{(2)} \\ x_2^{(2)} \\ x_3^{(2)} \end{bmatrix} \begin{bmatrix} 0 \\ 1 \\ 5 \end{bmatrix}$$

Hence, we have $x_3^{(2)} = \frac{1}{3}, \quad x_2^{(2)} = \frac{2}{3}, \quad x_1^{(2)} = -\frac{1}{3}$

Solve:

$$\begin{bmatrix} 3 & 2 & -1 \\ 0 & -\frac{1}{3} & \frac{11}{3} \\ 0 & 0 & 15 \end{bmatrix} \begin{bmatrix} x_1^{(3)} \\ x_2^{(3)} \\ x_3^{(3)} \end{bmatrix} = \begin{bmatrix} 0 \\ 0 \\ 1 \end{bmatrix}$$

Hence, we obtain $x_3^{(3)} = \frac{1}{15}, \quad x_2^{(3)} = \frac{11}{15}, \quad x_1^{(3)} = -\frac{7}{15}$

Thus, we have

$$\mathbf{A}^{-1} = \begin{bmatrix} \frac{11}{15} & -\frac{1}{3} & -\frac{7}{15} \\ -\frac{13}{15} & \frac{2}{3} & \frac{11}{15} \\ -\frac{8}{15} & \frac{1}{3} & \frac{1}{15} \end{bmatrix} = \frac{1}{15} \begin{bmatrix} 11 & -5 & -7 \\ -13 & 10 & 11 \\ -8 & 5 & 1 \end{bmatrix}$$

8.1.14. a.

$$\begin{bmatrix} a_{11} & a_{12} & 0 & 0 \\ a_{21} & a_{22} & a_{23} & 0 \\ 0 & a_{32} & a_{33} & a_{34} \\ 0 & 0 & a_{43} & a_{44} \end{bmatrix} = \begin{bmatrix} \ell_{11} & 0 & 0 & 0 \\ \ell_{21} & \ell_{22} & 0 & 0 \\ 0 & \ell_{32} & \ell_{33} & 0 \\ 0 & 0 & \ell_{43} & \ell_{44} \end{bmatrix} \begin{bmatrix} 1 & u_{12} & 0 & 0 \\ 0 & 1 & u_{23} & 0 \\ 0 & 0 & 1 & u_{34} \\ 0 & 0 & 0 & 1 \end{bmatrix}$$

$$= \begin{bmatrix} \ell_{11} & \ell_{11}u_{12} & 0 & 0 \\ \ell_{21} & \ell_{21}u_{12} + \ell_{22} & \ell_{22}u_{23} & 0 \\ 0 & \ell_{32} & \ell_{32}u_{23} + \ell_{33} & \ell_{33}u_{34} \\ 0 & 0 & \ell_{43} & \ell_{43}u_{34} + \ell_{44} \end{bmatrix}$$

8.1.16.

$$\begin{aligned}\mathcal{X}\mathcal{X}^{-1} &= \begin{bmatrix}\mathbf{A} & \mathbf{B}\\ \mathbf{C} & \mathbf{D}\end{bmatrix}\begin{bmatrix}\mathbf{I} & -\mathbf{A}^{-1}\mathbf{B}\\ \mathbf{0} & \mathbf{I}\end{bmatrix}\begin{bmatrix}\mathbf{A}^{-1} & \mathbf{0}\\ \mathbf{0} & (\mathbf{D}-\mathbf{C}\mathbf{A}^{-1}\mathbf{B})^{-1}\end{bmatrix}\begin{bmatrix}\mathbf{I} & \mathbf{0}\\ -\mathbf{C}\mathbf{A}^{-1} & \mathbf{I}\end{bmatrix}\\ &= \begin{bmatrix}\mathbf{A} & \mathbf{0}\\ \mathbf{C} & (\mathbf{D}-\mathbf{C}\mathbf{A}^{-1}\mathbf{B})\end{bmatrix}\begin{bmatrix}\mathbf{A}^{-1} & \mathbf{0}\\ \mathbf{0} & (\mathbf{D}-\mathbf{C}\mathbf{A}^{-1}\mathbf{B})^{-1}\end{bmatrix}\begin{bmatrix}\mathbf{I} & \mathbf{0}\\ -\mathbf{C}\mathbf{A}^{-1} & \mathbf{I}\end{bmatrix}\\ &= \begin{bmatrix}\mathbf{I} & \mathbf{0}\\ \mathbf{C}\mathbf{A}^{-1} & \mathbf{I}\end{bmatrix}\begin{bmatrix}\mathbf{I} & \mathbf{0}\\ -\mathbf{C}\mathbf{A}^{-1} & \mathbf{I}\end{bmatrix} = \begin{bmatrix}\mathbf{I} & \mathbf{0}\\ \mathbf{0} & \mathbf{I}\end{bmatrix}\end{aligned}$$

a.

$$\mathbf{A} = \begin{bmatrix}1 & 0\\ 0 & 1\end{bmatrix}, \quad \mathbf{B} = \begin{bmatrix}0 & 1\\ 1 & 0\end{bmatrix}, \quad \mathbf{C} = \begin{bmatrix}1 & 0\\ 0 & 0\end{bmatrix}, \quad \mathbf{D} = \begin{bmatrix}1 & 2\\ 0 & 1\end{bmatrix}$$

$$\begin{aligned}\mathcal{X}^{-1} &= \begin{bmatrix}1 & 0 & 0 & -1\\ 0 & 1 & -1 & 0\\ 0 & 0 & 1 & 0\\ 0 & 0 & 0 & 1\end{bmatrix}\begin{bmatrix}1 & 0 & 0 & 0\\ 0 & 1 & 0 & 0\\ 0 & 0 & 1 & -1\\ 0 & 0 & 0 & 1\end{bmatrix}\begin{bmatrix}1 & 0 & 0 & 0\\ 0 & 1 & 0 & 0\\ -1 & 0 & 1 & 0\\ 0 & 0 & 0 & 1\end{bmatrix}\\ &= \begin{bmatrix}1 & 0 & 0 & -1\\ 1 & 1 & -1 & 1\\ -1 & 0 & 1 & -1\\ 0 & 0 & 0 & 1\end{bmatrix}\end{aligned}$$

b.

$$\mathbf{A} = \begin{bmatrix}1 & 0 & 0\\ 0 & 1 & 0\\ 0 & 0 & 1\end{bmatrix}, \quad \mathbf{B} = \begin{bmatrix}1\\ 1\\ 1\end{bmatrix}, \quad \mathbf{C} = [1 \quad 1 \quad 1], \quad \mathbf{D} = [2]$$

$$\begin{aligned}\mathcal{X}^{-1} &= \begin{bmatrix}1 & 0 & 0 & -1\\ 0 & 1 & 0 & -1\\ 0 & 0 & 1 & -1\\ 0 & 0 & 0 & 1\end{bmatrix}\begin{bmatrix}1 & 0 & 0 & 0\\ 0 & 1 & 0 & 0\\ 0 & 0 & 1 & 0\\ 0 & 0 & 0 & -1\end{bmatrix}\begin{bmatrix}1 & 0 & 0 & 0\\ 0 & 1 & 0 & 0\\ 0 & 0 & 1 & 0\\ -1 & -1 & -1 & 1\end{bmatrix}\\ &= \begin{bmatrix}0 & -1 & -1 & 1\\ -1 & 0 & -1 & 1\\ -1 & -1 & 0 & 1\\ 1 & 1 & 1 & -1\end{bmatrix}\end{aligned}$$

8.1 Computer Exercises

8.1.C3.

$$\textbf{Case 4.} \quad \begin{bmatrix}536 & -668 & 458 & -186\\ -668 & 994 & -854 & 458\\ 458 & -854 & 994 & -668\\ -186 & 458 & -668 & 536\end{bmatrix}$$

8.2 Eigenvalues and Eigenvectors

8.2.1. Yes.

8.2.3. The bases vectors $(1,0)$ and $(0,1)$ map to $(\cos\theta, \sin\theta)$ and $(-\sin\theta, \cos\theta)$, respectively. From a sketch, it is easy o see that this is a counterclockwise rotation through the angle θ.

8.2.7. **a.** Eigenvalues: $9.8393, 3.0804 \pm 1.3763i$

c. Eigenvalues: $0.4660 - 1.4971i$, $2.6112 + 0.3313i$, $-0.0773 + 2.1658i$

8.2.9. Answer **c.**

$$\begin{aligned}(\mathbf{I} - s\,\mathbf{uu}^*)(\mathbf{I} - s\,\mathbf{uu}^*)^* &= (\mathbf{I} - s\,\mathbf{uu}^*)(\mathbf{I}^* - s\,\mathbf{uu}^*)\\ &= \mathbf{I} - s\,\mathbf{uu}^* - s\,\mathbf{uu}^* + s^2\,\mathbf{uu}^*\mathbf{uu}^*\\ &= \mathbf{I} - 2s\,\mathbf{uu}^* + s^2\,\mathbf{uu}^*\end{aligned}$$

So we want $-2s + s^2 = 0$, $s(-2+s) = 0$, $s = 0$ or $s = 2$.

8.2.11. Answer **d.**

$\mathbf{A}^T$ has the same eigenvalues as $\mathbf{A}$ because their characteristic polynomials are the same: $\text{Det}(\mathbf{A} - \lambda\mathbf{I}) = \text{Det}(\mathbf{A} - \lambda\mathbf{I})^T = \text{Det}(\mathbf{A}^T - \lambda\mathbf{I})$

8.2.13. True.

8.2.15. True.

8.2 Computer Exercises

8.2.C1. **a.** Eigenvalues: $-2 \pm \sqrt{23} = -6.79583, 2.79583$;
Eigenvectors: $[1, 9 - (3 \pm \sqrt{23})/7]^T$ or $[1, -1, 11369]^T$ and $[1, 0.25655]^T$.

d. $n = 5$: Eigenvalues: $0.26795, 1, 2, 3, 3.73205$

8.2.C7. Eigenvalues/Eigenvectors

a. $5.2426, (0.6656, 0.7463)$; $-3.2426, (-0.3051, 0.9523)$

b. $3.98923 \pm 5.5601i$, $(0.7267, -0.068 \pm 0.4533i, -0.2295 \mp 0,3829i)$;
$0.0241, (0.7916, 0.5137, 0.3308)$

8.2.C11. Answer **d.**

Eigenvalues, eigenvectors: $1, (-1,1,0,0)$; $2, (0,0,-1,1)$; $5, (-1,-1,2,2)$; $10, (2,2,1,1)$.

Eigenvalues/eigenvectors:

$$\mathbf{D} = \begin{bmatrix} 1.0000 & 0 & 0 & 0\\ 0 & 2.0000 & 0 & 0\\ 0 & 0 & 5.0000 & 0\\ 0 & 0 & 0 & 10.0000 \end{bmatrix},$$

$$\mathbf{U} = \begin{bmatrix} 0.7071 & -0.0000 & 0.3162 & 0.6325 \\ -0.7071 & -0.0000 & 0.3162 & 0.6325 \\ 0.0000 & 0.7071 & -0.6325 & 0.3162 \\ -0.0000 & -0.7071 & -0.6325 & 0.3162 \end{bmatrix}$$

8.3 Power Method

8.3.1. Answer **a**.

Because of the convergence, $\mathbf{x} = (h, t)^T$ should be an eigenvector.

The equation $\mathbf{Ax} = \lambda\mathbf{x}$ leads to $\lambda = 4h + 7$ and $\lambda h = 5h + 2$ or $(h + 1)(2h - 1) = 0$.

So either $h = -1$ or $h = \frac{1}{2}$.

Since the problems says $|h| < 1$, the answer is $h = \frac{1}{2}$.

Hence, we have $\lambda = 4h + 7 = 9$.

8.3.2. Answer **b**.

Pseudocode is power method applied to $\mathbf{A}^{-1}$.

Then $r = 1/\lambda$ should be the largest eigenvalue of $\mathbf{A}^{-1}$, where λ is the smallest eigenvalue of $\mathbf{A}$.

If $\mathbf{x}$ is an eigenvector satisfying $\mathbf{A}^{-1}\mathbf{x} = (1/\lambda)\mathbf{x}$, then $\mathbf{Xx} = \lambda\mathbf{x}$.

8.4 Iterative Solutions of Linear Systems

8.4.1. Matrix $\mathbf{A}$ is diaonally dominant, so the Jacobi and Gauss-Seidel methods converge.

Matrix $\mathbf{A}$ is symmetric and positive definite, so the SOR method converges.

Note: The eigenvalues of $\mathbf{A}$ are $(7 \pm \sqrt{33})/2 > 0$.

8.4.3. Answer **d**.

8.4.5. Answer **e**.

8.4.9. Answer **b**.

8.4 Computer Exercises

8.4.C1. Jacobi method: 77 iterations, $(-0.999962, 0.999956, -0.99999, 0.999954)$

Gauss-Seidel method: 38 iterations, $(-0.999957, 0.999956, -0.99992, 0.999957)$

SOR method: 12 iterations, $(-1.00003, 1.00002, -1.00001, 1.00000)$

8.4.C7. Both Jacobi and Gauss-Seidel methods converge to approximately the true solution $x = -4/11 \approx 0.\overline{36}$ and $y = 6/11 \approx 0.\overline{54}$

Chapter 9

Least Squares Method and Fourier Series

9.1 Method of Least Squares

9.1.1. Let $\varphi(c) = (c - \frac{5}{4})^2 + (c - \frac{4}{3})^2 + (c - \frac{5}{12})^2$. Set $0 = \frac{\partial \varphi}{\partial c} = 2(c - \frac{5}{4}) + 2(c - \frac{4}{3})$ $+2(c - \frac{5}{12}) = 2(3c - 3)$. Thus, we have $y(x) = 1$.

9.1.2. Let $\varphi(c) = \sum_{k=0}^{m}(c - y_k)^2$. Set $0 = \frac{\partial \varphi}{\partial c} = 2\sum_{k=0}^{m}(c - y_k) = 2\left[c(m+1) - \sum_{k=0}^{m} y_k\right]$.

Thus, we have $f(x) = \frac{1}{m+1}\sum_{k=0}^{m} y_k = (y_0 + \cdots + y_m)/(m+1)$, the *average* of the y values which does not involve any x_i.

9.1.3. Let $\varphi = \sum_{k=0}^{m}(ae^{x_k^2} + bx_k^3 - y_k)^2$. In particular, $\varphi = (ae - b)^2 + (a-1)^2 + (ae + b - 2)^2$.

Set $0 = \frac{\partial \varphi}{\partial a} = 2(ae - b)e + 2(a - 1) + 2(ae + b - 2)e$ and

$0 = \frac{\partial \varphi}{\partial b} = -2(ae - b) + 2(ae + b - 2)$. We have $a = (1 + 2e)/(1 + 2e^2)$ and $b = 1$.

9.1.5. Let $\varphi = \sum_{k=0}^{m}(ax_k^2 + b - y_k)^2$. In particular, we have
$\varphi = (a+b-3.1)^2 + (b-0.9)^2 + (a+b-2.9)^2$.
Set $0 = \dfrac{\partial\varphi}{\partial a} = 2(a+b-3.1) + 2(a+b-2.9)$ and
$0 = \dfrac{\partial\varphi}{\partial b} = 2(a+b-3.1) + 2(b-0.9) + 2(a+b-2.9)$.
Thus we have $\begin{cases} 2a + 2b = 6 \\ 2a + 3b = 6.9 \end{cases}$
and $a = 2.1$, $b = 0.9$. So $y = 2.1x^2 + 0.9$.

9.1.7. Let $\varphi = \sum_{k=0}^{m}(c\log x_k - y_k)^2$. Set $0 = \dfrac{\partial\varphi}{\partial c} = 2\sum_{k=0}^{m}(c\log x_k - y_k)(\log x_k)$. Hence, we have

$$c = \left(\sum_{k=0}^{m} y_k \log x_k\right) \Big/ \left[\sum_{k=0}^{m}(\log x_k)^2\right]$$

9.1.11. φ involves the sum of $m+1$ polynomials of degree two in c which is either concave upward or a constant. Thus, no maxima exists—only a minima.

9.1.12. Let $\varphi = \sum_{k=0}^{m}(\log(cx_k) - y_k)^2 = \sum_{k=0}^{m}(\log c + \log x_k - y_k)^2$. Set
$0 = \dfrac{\partial\varphi}{\partial c} = 2\sum_{k=0}^{m}(\log c + \log x_k - y_k)(\log e/c)$. Thus, we have
$(m+1)\log c = \sum_{k=0}^{m}(y_k - \log x_k)$ and $c = 10^{\left[(m+1)^{-1}\sum_{k=0}^{m}(y_k - \log x_k)\right]}$.

9.1.13. By (3) of Section 9.1, we have

$$\begin{bmatrix} 30 & 10 \\ 10 & 4 \end{bmatrix} \begin{bmatrix} a \\ b \end{bmatrix} = \begin{bmatrix} 13 \\ 4 \end{bmatrix}$$

and $a = 3/5$, $b = -1/2$. Thus, we have $y = (6x-5)/10$.

9.1.16. Let $\varphi = \sum_{k=0}^{m}(a + bT_k + cT_k^2 - V_k)^2$. Set $0 = \frac{\partial\varphi}{\partial a} = 2\sum_{k=0}^{m}(a + bT_k + cT_k^2 - V_k)$,

$0 = \frac{\partial\varphi}{\partial b} = 2\sum_{k=0}^{m}(a + bT_k + cT_k^2 - V_k)T_k$, $0 = 2\sum_{k=0}^{m}(a + bT_k + cT_k^2 - V_k)T_k^2$.

Thus, the normal equations are

$$\begin{bmatrix} m+1 & \sum_{k=0}^{m} T_k & \sum_{k=0}^{m} T_k^2 \\ \sum_{k=0}^{m} T_k & \sum_{k=0}^{m} T_k^2 & \sum_{k=0}^{m} T_k^3 \\ \sum_{k=0}^{m} T_k^2 & \sum_{k=0}^{m} T_k^3 & \sum_{k=0}^{m} T_k^4 \end{bmatrix} \begin{bmatrix} a \\ b \\ c \end{bmatrix} = \begin{bmatrix} \sum_{k=0}^{m} V_k \\ \sum_{k=0}^{m} V_kT_k \\ \sum_{k=0}^{m} V_kT_k^2 \end{bmatrix}$$

By the data, we have

$$\begin{bmatrix} 7 & 28 & 140 \\ 28 & 140 & 784 \\ 140 & 784 & 4676 \end{bmatrix} \begin{bmatrix} a \\ b \\ c \end{bmatrix} = \begin{bmatrix} 12.21 \\ 44.81 \\ 213.87 \end{bmatrix}$$

Thus, we have $a \approx 2.5929$, $b \approx -0.32583$, $c \approx 0.022738$.

9.1.18. Find the function of the form $f(x) = a\sin\pi x + b\cos\pi x$ that best fits the data in the text in the sense of least squares.

Let $\varphi = \sum_{k=0}^{m}(a\sin\pi x_k + b\cos\pi x_k - y_k)^2$.

We set

$$0 = \frac{\partial\varphi}{\partial a} = 2\sum_{k=0}^{m}(a\sin\pi x_k + b\cos\pi x_k - y_k)\sin\pi x_k$$

$$0 = \frac{\partial\varphi}{\partial b} = 2\sum_{k=0}^{m}(a\sin\pi x_k + b\cos\pi x_k - y_k)\cos\pi x_k$$

Hence, we obtain the normal equations are

$$\begin{bmatrix} \sum_{k=0}^{m}(\sin\pi x_k)^2 & \sum_{k=0}^{m}\sin\pi x_k\cos\pi x_k \\ \sum_{k=0}^{m}\sin\pi x_k\cos\pi x_k & \sum_{k=0}^{m}(\cos\pi x_k)^2 \end{bmatrix} \begin{bmatrix} a \\ b \end{bmatrix} = \begin{bmatrix} \sum_{k=0}^{m} y_k\sin\pi x_k \\ \sum_{k=0}^{m} y_k\cos\pi x_k \end{bmatrix}$$

We can then construct the following table

y_k	x_k	$\cos \pi x_k$	$(\cos \pi x_k)^2$	$\sin \pi x_k$	$(\sin \pi x_k)^2$	$\sin \pi x_k \cos \pi x_k$
-1	-1	-1	1	0	0	0
0	$-\frac{1}{2}$	0	0	-1	1	0
1	0	1	1	0	0	0
2	$\frac{1}{2}$	0	0	1	1	0
1	1	1	-1	1	0	0

The normal equations become

$$\begin{bmatrix} 2 & 0 \\ 0 & 3 \end{bmatrix} \begin{bmatrix} a \\ b \end{bmatrix} = \begin{bmatrix} 2 \\ 1 \end{bmatrix}$$

Hence, we obtain $a = 1$, $b = 1/3$, and $f(x) = \sin \pi x + (1/3) \cos \pi x$.

9.1.19. $y(x) = \sum_{j=0}^{m} c_j x^j$, the normal equations are

$$\begin{bmatrix} m+1 & \sum_{k=0}^{m} x_k & \sum_{k=0}^{m} x_k^2 & \cdots & \sum_{k=0}^{m} x_k^n \\ \sum_{k=0}^{m} x_k & \sum_{k=0}^{m} x_k^2 & & & \vdots \\ \sum_{k=0}^{m} x_k^2 & & \ddots & & \vdots \\ \vdots & & & \ddots & \vdots \\ \sum_{k=0}^{m} x_k^n & \cdots & \cdots & & \sum_{k=0}^{m} x_k^{2n} \end{bmatrix} \begin{bmatrix} c_0 \\ c_1 \\ c_2 \\ \vdots \\ c_n \end{bmatrix} = \begin{bmatrix} \sum_{k=0}^{m} y_k \\ \sum_{k=0}^{m} y_k x_k \\ \sum_{k=0}^{m} y_k x_k^2 \\ \vdots \\ \sum_{k=0}^{m} y_k x_k^n \end{bmatrix}$$

Here $m = 4$, $y = c_0 + c_1 x + c_2 x^2$, and

$$\begin{bmatrix} 5 & 0 & 10 \\ 0 & 10 & 0 \\ 10 & 0 & 34 \end{bmatrix} \begin{bmatrix} c_0 \\ c_1 \\ c_2 \end{bmatrix} = \begin{bmatrix} 7 \\ 0 \\ 18 \end{bmatrix}$$

Hence, we obtain $c_1 = 0$, $c_2 = 2/7$, $c_0 = 29/35$. Thus, we have $y(x) = (10x^2 + 29)/35$.

9.1.20.

$$d = 3(5) - 9 = 6$$

$$a = \tfrac{1}{6}[3(-6+14) - (3)(6)] = 1;$$

$$b = \tfrac{1}{6}[(5)(6) - (3)(8)] = 1$$

$$y = x + 1$$

9.1.21.

$$\varphi(c) = \sum_{k=0}^{m}[f(x_k) - ce^{x_k}]^2$$

$$\frac{\partial\varphi}{\partial c} = -2\sum_{k=0}^{m}[f(x_k) - ce^{x_k}]e^{x_k} = 0$$

$$c = \left[\sum_{k=0}^{m} e^{x_k} f(x_k)\right] \Big/ \left[\sum_{k=0}^{m} e^{2x_k}\right]$$

9.2 Orthogonal Systems and Chebyshev Polynomials

9.2.2. Let $w_k = c_k + 3xw_{k+1} + 2w_{k+2}$, $w_{n+2} = w_{n+1} = 0$. Now

$$\begin{aligned} f(x) &= \sum_{k=0}^{n} c_k g_k = \sum_{k=0}^{n} w_k g_k - 3x\sum_{k=0}^{n} w_{k+1} g_k - 2\sum_{k=0}^{n} w_{k+2} g_k \\ &= \sum_{j=0}^{n} w_j g_j - 3x\sum_{j=0}^{n} w_j g_{j-1} - 3xw_{n+1}g_n - 2\sum_{j=2}^{n} w_j g_{j-2} - w_{n+1}g_{n-1} - w_{n+2}g_n \\ &= w_0 g_0 + w_1 g_1 - 3xw_1 g_0 + \sum_{j=2}^{n} w_j (g_j - 3xg_{j-1} - 2g_{j-2}) \\ &= w_0 - w_1(1 + 2x) \end{aligned}$$

9.2.3. Since $\cos(n-2)\theta = \cos[(n-1)\theta - \theta] = \cos(n-1)\theta\cos\theta + \sin(n-1)\theta\sin\theta$, we have $2\cos\theta\cos(n-1)\theta - \cos(n-2)\theta = \cos(n-1)\theta\cos\theta - \sin(n-1)\theta\sin\theta = \cos(n\theta)$.
Note if $g_n(\theta) = \cos n\theta$, then $g_n(\theta) = 2\cos\theta g_{n-1}(\theta) - g_{n-2}(\theta)$.

9.2.5. By the previous exercise, the recursive relation is the same as (2) so that $T_n(x) = f_n(x) = \cos(n \arccos x)$.

9.2.6. $T_n(T_m(x)) = \cos(n\arccos(\cos(m\arccos x))) = \cos(nm\arccos x) = T_{nm}(x)$.

9.2.7. $|T_n(x)| = |\cos(n\arccos x)| \le 1$ for all $x \in [-1, 1]$ since $|\cos y| \le 1$ and for $\arccos x$ to exist x must be $|x| \le 1$.

9.2.8. Let $y = \frac{1}{2}x + \frac{1}{2}$. By (2), we have $T_j(y) = 2yT_{j-1}(y) - T_{j-2}(y)$ $= (x+1)T_{j-1}(y) - T_{j-2}(y)$. Hence, we obtain the recursive relation is

$$\begin{cases} g_0(x) = 1, \ g_1(x) = \frac{1}{2}x + \frac{1}{2} \\ g_j(x) = (x+1)g_{j-1}(x) - g_{j-2}(x) \end{cases}$$

9.2.10. $n + 2$ multiplications, $2n + 1$ additions/subtractions if $2x$ is computed as $x + x$.

9.2.12. n multiplications, $2n$ additions/subtractions.

9.2.13. Using $T_{n+1} = 2xT_n - T_{n-1}$, we obtain

$$
\begin{aligned}
T_0 &= 1, \quad T_1 = x \\
T_2 &= 2xT_1 - T_0 = 2x^2 - 1 \\
T_3 &= 2xT_2 - T_1 = 2x(2x^2 - 1) - x = 4x^3 - 3x \\
T_4 &= 2xT_3 - T_2 = 2x(4x^3 - 3x) - (2x^2 - 1) = 8x^4 - 8x^2 + 1 \\
T_5 &= 2xT_4 - T_3 = 2x(8x^4 - 8x^2 + 1) - (4x^3 - 3x) = 16x^5 - 20x^3 + 5x \\
T_6 &= 2xTT_5 - T_4 = 2x(16x^5 - 20x^3 + 5x) - (8x^4 - 8x^2 + 1) = 32x^6 - 48x^4 + 18x^2 - 1
\end{aligned}
$$

9.2.17. This is an interpolation problem for a table having two entries (x_1, y_1) and (x_2, y_2). The conditions on α and β are

$$
\begin{cases}
\alpha x_y 1^{12} + \beta x_1^{13} = y_1 \\
\alpha x_2^{12} + \beta x_2^{13} = y_2 .
\end{cases}
$$

Using determinants to solve this, we have

$$
\Delta = \begin{vmatrix} x_1^{12} & x_1^{13} \\ x_2^{12} & x_2^{13} \end{vmatrix} = x_1^{12} x_2^{12} \begin{vmatrix} 1 & x_1 \\ 1 & x_2 \end{vmatrix} = x_1^{12} x_2^{12} (x_2 - x_1) .
$$

Hence, we have

$$
\alpha = \begin{vmatrix} y_1 & x_1^{13} \\ y_2 & x_2^{13} \end{vmatrix} \div \Delta = \frac{y_1 x_2^{13} - y_2 x_1^{13}}{x_1^{12} x_2^{12} (x_2 - x_1)} .
$$

To see whether α is sensitive to changes in y_1, we compute $\partial\alpha/\partial y_1 = x_2^{13}/[x_1^{12} x_2^{12}(x_2 - x_1)] = x_2/[x_1^{12}(x_2 - x_1)]$.

In the given problem, this last quantity is $(0.9)/[10^{-12}(0.8)] \approx 10^{12}$. Thus, α is very sensitive to perturbations in y_1.

9.2 Computer Exercises

9.2.C7. $a_{ij} = \begin{cases} 0, & i \neq j \\ m+1, & i = j = 1 \\ (m+2)/2, & i = j \neq 1 \end{cases}$

9.3 Examples of the Least Square Principle

9.3.2. By (5), the coefficient matrix or the normal equations has elements

$$a_{ij} = \int_0^1 x^{i-1}x^{j-1}\,dx = \frac{x^{i+j-1}}{i+j-1}\bigg|_0^1 = \frac{1}{i+j-1}$$

which is the $(n+1)\times(n+1)$ Hilbert matrix.

9.3.3. Let $\varphi = \sum_{k=0}^{m}(e^{cx_k} - y_k)^2$. In particular, we have $\varphi = (1-\frac{1}{2})^2 + (e^c - 1)^2$. Set $0 = \dfrac{\partial\varphi}{\partial c} = 2(e^c - 1)e^c$. Hence, we have $e^c = 1$ or $c = 0$. Thus, we have $y = 1$.

9.3.4. Now $\varphi = (1-a)^2 + (e^c - b)^2$. Set $0 = \dfrac{\partial\varphi}{\partial c} = 2(e^c - b)e^c$. Hence, we have $e^c = b$ and $y = (e^c)^x = b^x$.

9.3.6. Let $\varphi = (e^c + 12)^2 + (e^{2c} - 15/2)^2$. Set $0 = \dfrac{\partial\varphi}{\partial c} = 2(e^c + 12)e^c + 2(e^{2c} - 15/2)e^{2c}2$. Hence, we obtain $2e^c(e^c + 12 + 2e^{3c} - 15e^c) = 0$. We have $e^c[(e^c)^3 - 7(e^c) + 6] = 0$. Hence, we obtain $e^c = 2$.

9.3.8. Let $\varphi = (2x + 3y - 1)^2 + (x - 4y + 9)^2 + (2x - y + 1)^2$. Set

$$0 = \frac{\partial\varphi}{\partial x} = 4(2x + 3y - 1) + 2(x - 4y + 9) + 4(2x - y + 1) \Rightarrow x = -1$$
$$0 = \frac{\partial\varphi}{\partial y} = 6(2x + 3y - 1) - 8(x - 4y + 9) - 2(2x - y + 1) \Rightarrow y = 20/13$$

9.3.9. **a.** Let $\varphi = \int_0^{\pi/2} [cx - \sin x]^2\,dx$. Set $0 = \dfrac{\partial\varphi}{\partial c} = 2\int_0^{\pi/2} (cx - \sin x)x\,dx$. Hence, we obtain $c\int_0^{\pi/2} x^2\,dx = \int_0^{\pi/2} x\sin x\,dx$ is the normal equation. So we have

$$c\frac{x^3}{3}\bigg|_0^{\pi/2} = \sin x - x\cos x\bigg|_0^{\pi/2} \Rightarrow c\Big(\frac{\pi^3}{24}\Big) = 1$$

Thus, we obtain $c = 24/\pi^3 \approx 0.77403\,68264$.

b. Let $\varphi = \int_0^1 [cx - e^x]^2\,dx$. Set $0 = \dfrac{\partial\varphi}{\partial c} = 2\int_0^1 (cx - e^x)(x)\,dx$. Hence, we have $c\int_0^1 x^2\,dx = \int_0^1 xe^x\,dx$ is the normal equation. So $c\left(\frac{1}{3}\right) = 1$ and $c = 3$.

9.3.14. We see that the functions are different and the solutions are different.

$$y \approx a + \frac{b}{x} + \frac{c}{x^2}; \sum_{i=0}^{m} (\text{residuals})^2 = \sum_{i=0}^{m} \left[y_i - a - \frac{b}{x_i} - \frac{c}{x^2} \right]^2 = \varphi(a, b, c)$$

$$yx^2 \approx ax^2 + bx + c; \sum_{i=0}^{m} (\text{residuals})^2 = \sum_{i=0}^{m} \left[y_i x_i^2 - a x_i^2 - b x_i - c \right]^2 = \psi(a, b, c)$$

$$= \sum_{i=0}^{m} \left[y_i - a - \frac{b}{x_i} - \frac{c}{x_i^2} \right]^2 x_i^4$$

9.3.15. $y \approx \dfrac{1}{a + bx}$. Change to $\dfrac{1}{y} \approx a + bx$. This is linear. Normal equations:

$$\varphi(a, b) = \sum_{i=0}^{m} \left(\frac{1}{y_i} - a - bx_i \right)^2; \qquad \frac{\partial \varphi}{\partial a} = \cdots = 0, \text{ etc.}$$

9.3.16. Let $\varphi = \int_0^\pi [a + b\cos x + c\sin x - e^{2x}]^2 \, dx$. By (5), we have $a_{ij} = \int_0^\pi g_i g_j \, dx$ where $g_0 = 1$, $g_1 = \cos x$, $g_2 = \sin x$. Now $a_{00} = \int_0^\pi dx = \pi$, $a_{01} = \int_0^\pi \cos x \, dx = 0$,
$a_{02} = \int_0^\pi \sin x \, dx = 2$, $a_{11} = \int_0^\pi \cos^2 x \, dx = \frac{1}{2}x + \frac{1}{4} \sin 2x\big|_0^\pi = \frac{\pi}{2}$,
$a_{12} = \int_0^\pi \cos x \sin x \, dx = \frac{1}{2} \sin^2 x\big|_0^\pi = 0$, $a_{22} = \int_0^\pi \sin^2 x \, dx = \frac{1}{2}x - \frac{1}{4} \sin^2 x\big|_0^\pi = \frac{\pi}{2}$.
Moreover, $b_i = \int_0^\pi f g_i \, dx$. So $b_0 = \int_0^\pi e^{2x} \, dx = \frac{1}{2} e^{2x}\big|_0^\pi = \frac{1}{2}(2^{2\pi} - 1)$,
$b_2 = \int_0^\pi e^{2x} \sin x \, dx = \frac{1}{5} e^{2x}(2\sin x - \cos x)\big|_0^\pi = \frac{1}{5}(e^{2\pi} + 1)$, $b_1 = \int_0^\pi e^{2x} \cos x \, dx = \frac{1}{5}$
$e^{2x}(2\cos x + \sin x)\big|_0^\pi = -\frac{2}{5}(e^{2\pi} + 1)$. The normal equations are

$$\begin{bmatrix} \pi & 0 & 2 \\ 0 & \pi/2 & 0 \\ 2 & 0 & \pi/2 \end{bmatrix} \begin{bmatrix} a \\ b \\ c \end{bmatrix} = \begin{bmatrix} \frac{1}{2}(e^{2\pi} - 1) \\ -\frac{2}{5}(e^{2\pi} + 1) \\ \frac{1}{5}(e^{2\pi} + 1) \end{bmatrix}$$

9.3.17. $\varphi(c) = \int_0^1 (e^x - cx)^2 \, dx$, $\dfrac{\partial \varphi}{\partial c} = -2 \int_0^1 x(e^x - cx) \, dx = 0$
$\int_0^1 xe^x \, dx = c \int_0^1 x^2 \, dx \Rightarrow e^x(x - 1)\big|_0^1 = c \dfrac{x^3}{3}\Big|_0^1 \Rightarrow c = 3$

9.3.20. Given $y = c\sin x$. We have $\sum_{i=0}^{m}[c\sin x_i - y_i]^2 = \min$. Differentiate wrt c so that

$$\sum_{i=0}^{m} 2\left[c\sin x_i - y_i\right]\sin x_i = 0$$

Hence, we have $c\sum_{i=0}^{m}(\sin x_i)^2 - \sum_{i=0}^{m} y_i \sin x_i = 0$. Therefore, we obtain

$$c = \left[\sum_{i=0}^{m} y_i \sin x_i\right] \Big/ \left[\sum_{i=0}^{m} (\sin x_i)^2\right]$$

9.3 Computer Exercise

9.3.C1. $a = 2$, $b = 3$.

9.4 Fourier Series

9.4.5. Consider $f(x) = x^k$, for $k \geq 0$, over $[-L, L]$.

a. **Case k odd:** Suppose that k is odd, then $f(x)$ is an odd function and all $a_n = 0$ for $n \geq 0$. First, let's check

$$\begin{aligned} a_0 &= \frac{1}{L}\int_{-L}^{L} f(x)\,dx = \frac{1}{L}\int_{-L}^{0} x^k\,dx + \frac{1}{L}\int_{0}^{L} x^k\,dx \\ &= \frac{1}{L}\left(\frac{x^{k+1}}{k+1}\right)\Big|_{-L}^{0} + \frac{1}{L}\left(\frac{x^{k+1}}{k+1}\right)\Big|_{0}^{L} \\ &= \frac{2}{k+1}(-L)^k + \frac{2}{k+1}L^k \\ &= 0 \end{aligned}$$

when k is odd. Next, we determine

$$\begin{aligned} b_n &= \frac{1}{L}\int_{-L}^{L} f(x)\sin\left(\frac{n\pi}{L}x\right) dx \\ &= \frac{2}{L}\int_{0}^{L} x^k \sin\left(\frac{n\pi}{L}x\right) dx \end{aligned}$$

Using integration by parts with $u = x^k$, $dv = \sin\big((n\pi/L)x\big)\,dx$, $du = kx^{k-1}\,dx$, $v = \int \sin\big((n\pi/L)x\big)\,dx = -\big((L/(n\pi)\big)\cos\big((n\pi/L)x\big)$, and $\int u\,dv = uv - \int v\,du$, we obtain

$$\begin{aligned}
b_n &= \frac{2}{L}\left[-x^k\left(\frac{L}{n\pi}\right)\cos\left(\frac{n\pi}{L}x\right)\Big|_0^L + \int_0^L kx^{k-1}\left(\frac{L}{n\pi}\right)\cos\left(\frac{n\pi}{L}x\right)dx\right] \\
&= -\left(\frac{2L^k}{n\pi}\right)\cos n\pi + \left(\frac{2k}{n\pi}\right)\int_0^L x^{k-1}\cos\left(\frac{n\pi}{L}x\right)dx \\
&= -(-1)^n\left(\frac{2L^k}{n\pi}\right) + \left(\frac{2k}{n\pi}\right)\left[\int_0^L x^{k-1}\cos\left(\frac{n\pi}{L}x\right)dx\right]
\end{aligned}$$

since $\cos n\pi = (-1)^n$. Again we use integration by parts with $u = x^{k-1}$, $dv = \cos\big((n\pi/L)x\big)\,dx$, $du = (k-1)x^{k-2}dx$, $v = \int \cos\big((n\pi/L)x\big)\,dx = (L/(n\pi))\sin\big((n\pi)/L)x\big)$, and $\int u\,dv = uv - \int v\,du$.
Now, we obtain

$$\begin{aligned}
b_n &= -(-1)^n\left(\frac{2L^k}{n\pi}\right) + \left(\frac{2k}{n\pi}\right)\left[x^{k-1}\left(\frac{L}{n\pi}\right)\sin\left(\frac{n\pi}{L}x\right)\Big|_0^L\right. \\
&\quad \left. - \int_0^L (k-1)x^{k-2}\left(\frac{L}{n\pi}\right)\sin\left(\frac{n\pi}{L}x\right)dx\right] \\
&= -(-1)^n\left(\frac{2L^k}{n\pi}\right) + \frac{2kL}{n^2\pi^2}\left[L^{k-1}\sin n\pi - (k-1)\int_0^L x^{k-2}\sin\left(\frac{n\pi}{L}x\right)dx\right] \\
&= -(-1)^n\left(\frac{2L^k}{n\pi}\right) + \frac{2kL}{n^2\pi^2}\left[-(k-1)\int_0^L x^{k-2}\sin\left(\frac{n\pi}{L}x\right)dx\right] \\
&= -(-1)^n\left(\frac{2L^k}{n\pi}\right) - \frac{k(k-1)L^2}{n^2\pi^2}\left[\frac{2}{L}\int_0^L x^{k-2}\sin\left(\frac{n\pi}{L}x\right)dx\right]
\end{aligned}$$

If $k \geq 3$ is odd, $k-2$ is also odd. After repeating integration by parts twice, we obtain

$$\begin{aligned}
b_n &= -(-1)^n\frac{2L^k}{n\pi} - \frac{k(k-1)L^2}{n^2\pi^2}\left[-(-1)^n\frac{2L^{k-2}}{n\pi}\right. \\
&\quad \left. - \frac{(k-2)(k-3)L^2}{n^2\pi^2}\left[\frac{2}{L}\int_0^L x^{k-4}\cos\left(\frac{n\pi}{L}x\right)dx\right]\right] \\
&= -(-1)^n\frac{2L^k}{n\pi} + (-1)^n\frac{2k(k-1)L^k}{n^3\pi^3} + \frac{k(k-1)(k-2)(k-3)L^4}{n^4\pi^4}\left[\cdots\right]
\end{aligned}$$

If $k = 1$, we obtain

$$b_n = -(-1)^n\frac{2L}{n\pi}$$

and when $L = \pi$, we have

$$b_n = -(-1)^n \frac{2}{n}$$

If $k = 3$, we obtain

$$\begin{aligned} b_n &= -(-1)^n \frac{2L^3}{n\pi} + (-1)^n \frac{12L^3}{n^3\pi^3} \\ &= -(-1)^n \frac{2L^3}{n^3\pi^3}(-6 + n^2\pi^2) \end{aligned}$$

and when $L = \pi$, we have

$$b_n = -(-1)^n \frac{2}{n^3}(-6 + n^2\pi^2)$$

Hence, the Fourier series expansion for $f(x) = x$ over $[-L, L]$ is

$$x = -\frac{2L}{\pi} \sum_{n=1}^{\infty} (-1)^n \frac{1}{n} \sin\left(\frac{n\pi}{L} x\right)$$

and over $[-\pi, \pi]$ it is

$$x = -2 \sum_{n=1}^{\infty} (-1)^n \frac{1}{n} \sin nx$$

Moreover, the Fourier series expansion for $f(x) = x^3$ over $[-L, L]$ is

$$x^3 = -\frac{2L^3}{\pi^3} \sum_{n=1}^{\infty} (-1)^n \frac{1}{n^3}(-6 + n^2\pi^2) \sin\left(\frac{n\pi}{L} x\right)$$

and over $[-\pi, \pi]$ it is

$$x^3 = -2 \sum_{n=1}^{\infty} (-1)^n \frac{1}{n^3}(-6 + n^2\pi^2) \sin nx$$

b. Case k even: Suppose that k is even, then $f(x)$ is an even function and all $b_n = 0$ for $n \geq 1$. First, we find

$$a_0 = \frac{1}{L} \int_{-L}^{L} f(x)\, dx = \frac{2}{L} \int_0^L x^k\, dx = \frac{2}{L} \left(\frac{x^{k+1}}{k+1} \right) \Big|_0^L = \frac{2}{k+1} L^k$$

Next, we determine

$$\begin{aligned} a_n &= \frac{1}{L} \int_{-L}^{L} f(x) \cos\left(\frac{n\pi}{L} x\right) dx \\ &= \frac{2}{L} \int_0^L x^k \cos\left(\frac{n\pi}{L} x\right) dx \end{aligned}$$

Using integration by parts with $u = x^k$, $dv = \cos\big((n\pi/L)x\big)\,dx$, $du = kx^{k-1}\,dx$, $v = \int \cos\big((n\pi/L)x\big)\,dx = \big((L/(n\pi))\sin\big((n\pi/L)x\big)$, and $\int u\,dv = uv - \int v\,du$, we obtain

$$\begin{aligned} a_n &= \frac{2}{L}\left[x^k\left(\frac{L}{n\pi}\right)\sin\left(\frac{n\pi}{L}x\right)\Big|_0^L - \int_0^L kx^{k-1}\left(\frac{L}{n\pi}\right)\sin\left(\frac{n\pi}{L}x\right)dx\right] \\ &= \left(\frac{2L^k}{n\pi}\right)\sin n\pi - \left(\frac{2k}{n\pi}\right)\int_0^L x^{k-1}\sin\left(\frac{n\pi}{L}x\right)dx \\ &= -\left(\frac{2k}{n\pi}\right)\left[\int_0^L x^{k-1}\sin\left(\frac{n\pi}{L}x\right)dx\right] \end{aligned}$$

since $\sin n\pi = 0$. Again we use integration by parts with $u = x^{k-1}$, $dv = \sin\big((n\pi/L)x\big)\,dx$, $du = (k-1)x^{k-2}dx$, $v = \int \sin\big((n\pi/L)x\big)\,dx = -(L/(n\pi))\cos\big((n\pi)/L)x\big)$, and $\int u\,dv = uv - \int v\,du$.
Now, we obtain

$$\begin{aligned} a_n &= -\left(\frac{2k}{n\pi}\right)\left[-x^{k-1}\left(\frac{L}{n\pi}\right)\cos\left(\frac{n\pi}{L}x\right)\Big|_0^L + \int_0^L (k-1)x^{k-2}\left(\frac{L}{n\pi}\right)\cos\left(\frac{n\pi}{L}x\right)dx\right] \\ &= \frac{2kL}{n^2\pi^2}\left[L^{k-1}\cos n\pi - (k-1)\int_0^L x^{k-2}\cos\left(\frac{n\pi}{L}x\right)dx\right] \\ &= \frac{2kL}{n^2\pi^2}\left[(-1)^n L^{k-1} - (k-1)\int_0^L x^{k-2}\cos\left(\frac{n\pi}{L}x\right)dx\right] \\ &= (-1)^n\frac{2kL^k}{n^2\pi^2} - \frac{k(k-1)L^2}{n^2\pi^2}\left[\frac{2}{L}\int_0^L x^{k-2}\cos\left(\frac{n\pi}{L}x\right)dx\right] \end{aligned}$$

If $k \geq 2$ is even, $k-2$ is also even. After repeating integration by parts twice, we obtain

$$\begin{aligned} a_n &= (-1)^n\frac{2kL^k}{n^2\pi^2} - \frac{k(k-1)L^2}{n^2\pi^2}\left[(-1)^n\frac{2(k-2)L^{k-2}}{n^2\pi^2}\right. \\ &\qquad \left. - \frac{(k-2)(k-3)L^2}{n^2\pi^2}\left[\frac{2}{L}\int_0^L x^{k-4}\cos\left(\frac{n\pi}{L}x\right)dx\right]\right] \\ &= (-1)^n\frac{2kL^k}{n^2\pi^2} - (-1)^n\frac{2k(k-1)(k-2)L^k}{n^4\pi^4} \\ &\qquad + \frac{k(k-1)(k-2)(k-3)L^4}{n^4\pi^4}\left[\cdots\right] \end{aligned}$$

If $k = 2$, we obtain

$$\frac{1}{2}a_0 = \frac{1}{3}L^2, \qquad a_n = (-1)^n\frac{4L^2}{n^2\pi^2} \quad (n \geq 1)$$

and when $L = \pi$, we have

$$\frac{1}{2}a_0 = \frac{1}{3}\pi^2, \qquad a_n = (-1)^n\frac{4}{n^2} \quad (n \geq 1)$$

If $k = 4$, we obtain $\frac{1}{2}a_0 = \frac{1}{5}L^4$ and, for $n \geq 1$,

$$\begin{aligned} a_n &= (-1)^n \frac{8L^4}{n^2\pi^2} - (-1)^n \frac{48L^4}{n^4\pi^4} + \frac{24L^4}{n^4\pi^4}\left[\frac{2}{L}\int_0^L \cos\left(\frac{n\pi}{L}x\right)dx\right] \\ &= (-1)^n \frac{8L^4}{n^2\pi^2} - (-1)^n \frac{48L^4}{n^4\pi^4} + \frac{24L^4}{n^4\pi^4}\left[\left(\frac{L}{n\pi}\right)\sin\left(\frac{n\pi}{L}x\right)\Big|_0^L\right] \\ &= (-1)^n \frac{8L^4}{n^2\pi^2} - (-1)^n \frac{48L^4}{n^4\pi^4} + \frac{24L^4}{n^4\pi^4}\sin n\pi \\ &= (-1)^n \frac{8L^4}{n^4\pi^4}(-6 + n^2\pi^2) \end{aligned}$$

and when $L = \pi$, we have

$$\frac{1}{2}a_0 = \frac{\pi^4}{5}, \qquad a_n = (-1)^n \frac{8i}{n^4}(-6 + n^2\pi^2)$$

The Fourier series expansion for the parabolic wave $f(x) = x^2$ over $[-L, L]$ is

$$x^2 = \frac{1}{3}L^2 + 4\frac{L^2}{\pi^2}\sum_{n=1}^{\infty}(-1)^n \frac{1}{n^2}\cos\left(\frac{n\pi}{L}x\right)$$

and over $[-\pi, \pi]$ it is

$$x^2 = \frac{\pi^2}{3} + 4\sum_{n=1}^{\infty}(-1)^n \frac{1}{n^2}\cos nx$$

Moreover, the Fourier series expansion for $f(x) = x^4$ over $[-L, L]$ it is

$$x^4 = \frac{1}{5}L^4 + 8\frac{L^4}{\pi^4}\sum_{n=1}^{\infty}(-1)^n \frac{1}{n^4}(-6 + n^2 n^2)\cos\left(\frac{n\pi}{L}x\right)$$

and over $[-\pi, \pi]$ is

$$x^4 = \frac{\pi^4}{5} + 8\sum_{n=1}^{\infty}(-1)^n \frac{1}{n^4}(-6 + n^2\pi^2)\cos nx$$

9.4.7. **d.** Consider the 2π-periodic function over $[-\pi, \pi]$, defined as $s(x) = -1$ on $(-\pi, -\pi/2)$, as $s(x) = 0$ on $(-\pi/, \pi/2)$, and as $s(x) = 1$ on $(-\pi/2, -\pi)$. Clearly, $s(x + 2\pi) = s(x)$ and the period is 2π. First, we have

$$\begin{aligned}
a_0 &= \frac{1}{\pi}\int_{-\pi}^{\pi} s(x)\,dx \\
&= \frac{1}{\pi}\Big[\int_{-\pi}^{-\pi/2} -1\,dx + \int_{-\pi/2}^{\pi/2} 0\,dx + \int_{\pi/2}^{\pi} 1\,dx\Big] \\
&= \frac{1}{\pi}\Big[-x\Big]_{-\pi}^{-\pi/2} + \frac{1}{\pi}\Big[x\Big]_{\pi/2}^{\pi} \\
&= \frac{1}{\pi}\Big[\frac{\pi}{2} - \pi + \pi - \frac{\pi}{2}\Big] \\
&= \frac{1}{\pi}\Big[-\frac{\pi}{2} + \frac{\pi}{2}\Big] = 0
\end{aligned}$$

Now we compute the other a_n coefficients, for $n \geq 1$,

$$\begin{aligned}
a_n &= \frac{1}{\pi}\int_{-\pi}^{\pi} s(x)\,dx \\
&= \frac{1}{\pi}\Big[\int_{-\pi}^{-\pi/2} -\cos nx\,dx + \int_{-\pi/2}^{\pi/2} 0\,dx + \int_{\pi/2}^{\pi} \cos nx\,dx\Big] \\
&= \frac{1}{\pi}\Big[-\frac{1}{n}\sin nx\Big|_{-\pi}^{-\pi/2} + \frac{1}{n}\sin nx\Big|_{\pi/2}^{\pi}\Big] \\
&= \frac{1}{n\pi}\Big[-\sin\big(-\frac{n\pi}{2}\big) + \sin\big(-n\pi\big) + \sin n\pi - \sin\frac{n\pi}{2}\Big] \\
&= \frac{1}{n\pi}\Big[\sin\frac{n\pi}{2} - \sin n\pi + \sin n\pi - \sin\frac{n\pi}{2}\Big] \\
&= 0
\end{aligned}$$

Notice that these results are obvious since $s(x)$ is an odd function! Now we find

$$\begin{aligned}
b_n &= \frac{1}{\pi}\int_{-\pi}^{\pi} s(x)\sin nx\,dx \\
&= \frac{1}{\pi}\Big[\int_{-\pi}^{-\pi/2} -\sin nx\,dx + \int_{-\pi/2}^{\pi} \sin nx\,dx\Big] \\
&= \frac{1}{n\pi}\Big[\cos(-\frac{n\pi}{2}) - \cos(-n\pi) - \cos n\pi + \cos\frac{n\pi}{2}\Big] \\
&= \frac{1}{n\pi}\Big[\cos\frac{nx}{2} - \cos n\pi - \cos n\pi + \cos\frac{n\pi}{2}\Big] \\
&= \frac{2}{n\pi}\Bigg[\cos\frac{n\pi}{2} - \cos n\pi\Bigg]
\end{aligned}$$

Thus, the Fourier series is

$$s(x) = \frac{2}{\pi}\sum_{n=1}^{\infty}\left(\cos\frac{n\pi}{2} - \cos n\pi\right)\sin n\pi$$

9.4.11. **a.** Write a in polar form

$$a = r(\cos\theta + i\sin\theta)$$

and use the de Moivre's Theorem

$$(\cos\theta + i\sin\theta)^n = \cos n\theta + i\sin n\theta$$

We obtain the n-th roots of a as

$$\sqrt[n]{a} = \sqrt[n]{r}\left[\cos\left(2\pi\frac{k}{n}\right) + i\sin\left(2\pi\frac{k}{n}\right)\right]$$

where $0 \leq k \leq n-1$.

b. We have $\sqrt[4]{2}$, $i\sqrt[4]{2}$, $-i\sqrt[4]{2}$, and $-\sqrt[4]{2}$. Note if $1, \omega, \omega^2, \ldots, \omega^{k-1}$ are the n-th roots of unity, then $\sqrt[n]{r}$, $\omega\sqrt[n]{r}$, ..., $\omega^{n-1}\sqrt[n]{r}$ are the n-th roots of $a = r(\cos\theta + i\sin\theta)$.

9.4.17. Recall that DFT for $n = 4$ is

$$y_m = \sum_{k=0}^{3}\omega_n^{mk}x_k$$

for $m = 0, 1, 2, 3$. In detail, we have

$$\begin{aligned}
y_0 &= \omega_4^0 x_0 + \omega_4^0 x_1 + \omega_4^2 x_2 + \omega_4^0 x_3\\
y_1 &= \omega_4^0 x_0 + \omega_4^1 x_1 + \omega_4^2 x_2 + \omega_4^3 x_3\\
y_2 &= \omega_4^0 x_0 + \omega_4^2 x_1 + \omega_4^4 x_2 + \omega_4^6 x_3\\
y_3 &= \omega_4^0 x_0 + \omega_4^3 x_1 + \omega_4^6 x_2 + \omega_4^9 x_3
\end{aligned}$$

Since $\omega_4^0 = 1 = \omega_4^4$, $\omega_4^2 = -1 = \omega_4^6$, $\omega_4^3 = i$, and $\omega_4^9 = -i = \omega_4^9$, we obtain

$$\begin{aligned}
y_0 &= (x_0 + x_2) + (x_1 + x_3)\\
y_1 &= (x_0 - x_2) + i(x_1 - x_3)\\
y_2 &= (x_0 + x_2) - (x_1 + x_3)\\
y_3 &= (x_0 - x_2) + i(x_1 - x_3)
\end{aligned}$$

We have reduced the number of operations from twelve additions and sixteen multiplications to six additions and two multiplications, at most.

9.4 Computer Exercises

9.4.C7. **b.** $g(x) = x(x+1)$ is a 2π-periodic function over $-\pi < x < \pi$ with $g(x+2\pi) = g(x)$ for all x. This function is neither even nor odd. It is 2π-periodic with $P = 2\pi$. First, we compute

$$\begin{aligned} a_0 &= \frac{1}{\pi}\int_{-\pi}^{\pi} g(x)\,dx = \frac{1}{\pi}\int_{-\pi}^{\pi}(x^2+x)\,dx \\ &= \frac{1}{\pi}\Big[\frac{x^3}{3}+\frac{x^2}{2}\Big]_{-\pi}^{\pi} = \frac{2\pi^2}{3} \end{aligned}$$

Next, we compute a_n, for $n \neq 0$,

$$\begin{aligned} a_n &= \frac{1}{\pi}\int_{-\pi}^{\pi} g(x)\cos nx\,dx \\ &= \frac{1}{\pi}\int_{-\pi}^{\pi} x(x+1)\cos nx\,dx \\ &= \frac{1}{\pi}\Big\{\Big[(x^2+x)\frac{\sin nx}{n}\Big]_{-\pi}^{\pi} - \int_{-\pi}^{\pi}(2x+1)\frac{\sin nx}{n}\,dx\Big\} \\ &= \frac{1}{n\pi}\int_{-\pi}^{\pi}(2x+1)\sin nx\,dx \\ &= \Big\{\Big[-(2x+1)\frac{\cos nx}{n}\Big]_{-\pi}^{\pi} + \int_{-\pi}^{\pi}\frac{2\cos nx}{n}\,dx\Big\} \\ &= -\frac{1}{n\pi}\Big\{-\frac{(2\pi+1)(-1)^n}{n} + \frac{(2(-\pi)+1)(-1)^n}{n} + \frac{2}{n}\Big[\frac{\sin nx}{n}\Big]_{-\pi}^{\pi}\Big\} \\ &= -\frac{1}{n\pi}\Big[-\frac{4\pi}{n}(-1)^n\Big] \\ &= (-1)^n\frac{4}{n^2} \end{aligned}$$

Finally, we compute

$$\begin{aligned}
b_n &= \frac{1}{\pi}\int_{-\pi}^{\pi}(x^2+x)\sin nx, dx\\
&= \frac{1}{\pi}\Big\{\Big[-(x^2+x)\frac{\cos nx}{n}\Big]_{-\pi}^{\pi} + \int_{-\pi}^{\pi}(2x+1)\frac{\cos nx}{n}\,dx\Big\}\\
&= \Big[-(\pi^2+\pi)\frac{(-1)^n}{n} + (\pi^2-\pi)\frac{(-1)^n}{n} + \frac{1}{n}\int_{-\pi}^{\pi}(2x+1)\cos n\pi\,dx\Big]\Big\}\\
&= \frac{1}{\pi}\Big[-2\pi\frac{(-1)^n}{n} + \frac{1}{n}\Big\{\Big[(2x+1)\frac{\sin nx}{n}\Big]_{-\pi}^{\pi} - \int_{-\pi}^{\pi}\frac{2\sin nx}{n}\,dx\Big]\Big\}\\
&= \frac{1}{\pi}\Big[-2\pi\frac{(-1)^n}{n} - \frac{2}{n^2}\int_{-\pi}^{\pi}2\sin nx\,dx\\
&= -(-1)^n\frac{2}{n}
\end{aligned}$$

Putting these formulas together, we obtain

$$x(x+1) = \frac{\pi^2}{3} + (-1)^n\sum_{n=1}^{\infty}\Big[\frac{4}{n^2}\cos nx - \frac{2}{n}\sin nx\Big]$$

Mathematica:

FourierCoefficient$[x(x+1), x, n]$
FourierSeries$[x(x+1), x, 2]$
FourierTrigSeries$[x(x+1), x, 2]$

$$\frac{(-1)^n(2+in)}{n^2}$$

$$(-2+i)e^{-ix} - (2+i)e^{ix} + \left(\frac{1}{2}-\frac{i}{2}\right)e^{-2ix} + \left(\frac{1}{2}+\frac{i}{2}\right)e^{2ix} + \frac{\pi^2}{3}$$

$$\frac{\pi^2}{3} + 4\left(-\text{Cos}[x] + \frac{1}{4}\text{Cos}[2x]\right) - 2\left(-\text{Sin}[x] + \frac{1}{2}\text{Sin}[2x]\right)$$

9.4.C8. We can find the Fourier series for the function $f(x) = x^k$, for $k = 1, 2, 3$, and 4, on the interval $[-\pi, \pi]$ using Mathematica:

a.

```
f[x_] := x
a[0] := (1/Pi) Integrate[f[x], {x,-Pi,Pi}]
a[0]
a[n_] := (1/Pi) Integrate[f[x]*Cos[n x], {x,-Pi,Pi}]
a[n]
TrigId = {Cos[Pi n_] -> (-1)^n, Sin[Pi n_] ->0};
b[n_] := (1/Pi) Integrate[f[x]*Sin[n x], {x,-Pi,Pi}]
b[n] /. TrigId
(* a[0]/2 + Sum[ a[n] Cos[n x] + b[n] Sin[n x],{n,1,N}] *)
Clear[FullSeries]
FullSeries[x_,N_]:= Sum[((2*(-1)^(n+1))/(n))Sin[2 Pi n x/(2*Pi)],{n,1,N}]
Plot[{FullSeries[x,9], f[x]}, {x,-Pi,Pi}]
```

$$x = -2\sum_{n=1}^{\infty}(-1)^n \frac{1}{n} \sin nx$$

b.

```
f[x_] := x^2
a[0] := (1/Pi) Integrate[f[x], {x,-Pi,Pi}]
a[0]
TrigId = {Cos[Pi n_] -> (-1)^n, Sin[Pi n_] ->0}
a[n_] := (1/Pi) Integrate[f[x]*Cos[n x], {x,-Pi,Pi}]
a[n] /. TrigId
b[n_] := (1/Pi) Integrate[f[x]*Sin[n x], {x,-Pi,Pi}]
b[n]
(* a[0]/2 + Sum[ a[n] Cos[n x] + b[n] Sin[n x],{n,1,N}]*)
Clear[FullSeries]
FullSeries[x_,N_]:=(Pi^2/3) + Sum[((4*(-1)^n)/(n^2))Cos[2 Pi n x/(2*Pi)],{n,1,N}]
Plot[{FullSeries[x,9], f[x]}, {x,-Pi,Pi}]
```

$$x^2 = \frac{\pi^2}{3} + 4\sum_{n=1}^{\infty}(-1)^n \frac{1}{n^2} \cos nx$$

c.

```
f[x_] := x^3
a[0] := (1/Pi) Integrate[f[x], {x,-Pi,Pi}]
a[0]
a[m_] := (1/Pi) Integrate[f[x]*Cos[n x], {x,-Pi,Pi}]
a[m]
TrigId = {Cos[Pi n_] -> (-1)^n, Sin[Pi n_] ->0}
b[n_] := (1/Pi) Integrate[f[x]*Sin[n x], {x,-Pi,Pi}]
b[n] /. TrigId
(* a[0]/2 + Sum[ a[n] Cos[n x] + b[n] Sin[n x],{n,1,N}]*)
Clear[FullSeries]
FullSeries[x_,N_]:=
Sum[((2*(-1)^n)(6-n^2*Pi^2)/(n^3))Sin[2 Pi n x/(2*Pi)],{n,1,N}]
Plot[{FullSeries[x,9], f[x]}, {x,-Pi,Pi}]
```

$$x^3 - = 2\sum_{n=1}^{\infty}(-1)^n\frac{1}{n^3}\left(-6+n^2\pi^2\right)\sin n\pi$$

d.

```
f[x_] := x^4
a[0] := (1/Pi) Integrate[f[x], {x,-Pi,Pi}]
a[0]
TrigId = {Cos[Pi n_] -> (-1)^n, Sin[Pi n_] ->0}
a[n_] := (1/Pi) Integrate[f[x]*Cos[n x], {x,-Pi,Pi}]
Simplify[a[n]] /.TrigId
b[n_] := (1/Pi) Integrate[f[x]*Sin[n x], {x,-Pi,Pi}]
b[n]
(* a[0]/2 + Sum[ a[n] Cos[n x] + b[n] Sin[n x],{n,1,N}]*)
Clear[FullSeries]
FullSeries[x_,N_]:= (Pi^4/5) + \
            Sum[((8*(-1)^m)(-6+m^2*Pi^2)/(m^4))Cos[2 Pi m x/(2*Pi)],{m,1,N}]
Plot[{FullSeries[x,9], f[x]}, {x,-Pi,Pi}]
```

$$x^4 = \frac{\pi^4}{5} + 8\sum_{n=1}^{\infty}(-1)^n\frac{1}{n^4}(-6+n^2\pi^2)\cos nx$$

Symbolic Computations of Fourier Coefficients

Mathematical software systems such as Mathematica, Maple, and MATLAB allow for symbolic computations and make it easy to determine the Fourier coefficients for a given function $f(x)$. When the function $f(x)$ is defined *piecewise*, we must *split* the integral accordingly with regard to the different intervals involved.

For example, consider the 2π-periodic **square wave** function over $(-\pi, \pi)$

$$f(x) = \begin{cases} -1, & -\pi < x < 0 \\ 1, & 0 < x < \pi \end{cases}$$

has the Fourier series

$$f(x) = \frac{4}{\pi} \sum_{n=1,3,5,\ldots}^{\infty} \frac{1}{n} \sin nx$$

Note that the function $f(x)$ is defined by two different formulas over two different intervals. So each Fourier coefficient integral from $-\pi$ to π must be calculated as the sum of two integrals as follows

$$a_n = \frac{1}{\pi} \int_{-\pi}^{0} -\cos nx\, dx + \frac{1}{\pi} \int_{0}^{\pi} \cos nx\, dx$$

$$b_n = \frac{1}{\pi} \int_{-\pi}^{0} -\sin nx\, dx + \frac{1}{\pi} \int_{0}^{\pi} \sin nx\, dx$$

Mathematica

First, we compute the cosine coefficients a_n

```
a[n_] = (1/Pi)*(Integrate[-Cos[n*x], {x, -Pi, 0}] +
                Integrate[ Cos[n*x], {x,   0, Pi}])
```

We obtain 0 for a_n. since our odd function $f(x)$ has no cosine terms in the Fourier series. Next, we determine the sine coefficients b_n

```
b[n_] = (1/Pi)*(Integrate[-Sin[n*x], {x, -Pi, 0}] +
                Integrate[ Sin[n*x], {x,   0, Pi}])
b[n] // Simplify
```

We obtain

$$b_n = -\frac{2}{n\pi}[1 - \cos(n\pi)] = \frac{2}{n\pi}[1 - (-1)^n] = \begin{cases} 4/n\pi, & n \text{ odd} \\ 0, & n \text{ even} \end{cases}$$

Finally, we compute some typical Fourier summations and plot its graph

```
fourierSum := Sum[b[n] Sin[n x], {n, 1, 9}]
plot(fourierSum, {x,-2 Pi, 4 Pi}];
```

From the output, we obtain

$$\frac{4}{\pi}\sin(x) + \frac{4}{3\pi}\sin(3x) + \frac{4}{5\pi}\sin(5x) + \frac{4}{7\pi}\sin(7x) + \frac{4}{9\pi}\sin(9x)$$

and the plot of its graph.

Maple

First, we compute the cosine coefficients a_n

```
a:= n -> (1/Pi)*(int(-cos(n*x), x=-Pi..0) + int(cos(n*x), x=0..Pi)):
a(n);
```

We obtain 0 for `a(n)` since our odd function has no cosine terms in the Fourier series. Next, we determine the sine coefficients b_n

```
b := n -> (1/Pi)*(int(-sin(n*x), x=-Pi..0) + int(sin(n*x), x=0..Pi)):
b(n);
```

We obtain

$$b_n = -2\Big(\frac{\cos(\pi n) - 1}{\pi n}\Big) = \frac{2}{n\pi}[1 - (-1)^n] = \begin{cases} 4/n\pi, & n \text{ odd} \\ 0, & n \text{ even} \end{cases}$$

Finally, we compute some typical Fourier summations and plot its graph

```
fourierSum := sum('b(n)*sin(n*x)', 'n'=1..9);
plot(fourierSum, x=-2*Pi..2*Pi);
```

From the output, we obtain for `fourierSum`

$$\frac{4}{\pi}\sin(x) + \frac{4}{3\pi}\sin(3x) + \frac{4}{5\pi}\sin(5x) + \frac{4}{7\pi}\sin(7x) + \frac{4}{9\pi}\sin(9x)$$

and the plot of its graph.

MATLAB

First, we determine the cosine coefficients a_n

```
syms n x pi
an = (1/pi)*(int(-cos(n*x), -pi, 0) + int(cos(n*x), 0, pi))
```

We obtain 0 for `an` since our odd function $f(x)$ has no cosine terms in the Fourier series. Next, we determine the sine coefficients b_n

```
bn = (1/pi)*(int(-sin(n*x), -pi, 0) + int(sin(n*x), 0, pi));
pretty(bn)
```

We obtain

$$b_n = -\frac{2}{n\pi}[\cos(\pi n) - 1] + 2\sin(\pi n)/2)^2)/pi * n)$$

Identities $\cos n\pi = (-1)^n$, $\sin[(n\pi)/2] = 0$ if n is even, and $\sin[n\pi)/2] = (-1)^n$ if n is odd are useful in simplifying this expression when using MATLAB commands such asy

```
bn = subs(bn, 'cos(pi*n)','(-1)^n'));
pretty(bn)
```

We obtain

$$b_n = \begin{cases} 4/n\pi, & n \text{ odd} \\ 0, & n \text{ even} \end{cases}$$

Hence, we obtain

$$b_n = \begin{cases} 4/n\pi, & n \text{ odd} \\ 0, & n \text{ even} \end{cases}$$

Finally, we compute some typical Fourier summations and plot theirs graphs

```
FourierSum = (4/pi)*sin(x);
for k = 3:2:9
   FourierSum = FourierSum + subs((4/pi)*sin(n*x)/n, k, n);
end
FourierSum
ezplot(FourierSum, 3.1416*[-2\quad 4])
```

We obtain this partial output

$$\frac{4}{\pi}\sin(x) + \frac{4}{3\pi}\sin(3x) + \frac{4}{5\pi}\sin(5x) + \frac{4}{7\pi}\sin(7x) + \frac{4}{9\pi}\sin(9x)$$

and the plot of its graph.

MATLAB Example Code

```
% Compute Fourier Series (MATLAB)
% f(x) = x

% Set-up for n-th partial sums [-L,L]
syms x k L n

% Set-up k is an integer
evalin(symengine, 'assume(k, Type::Integer)');

% Define k-th cosine coefficient
a = @(f,x,k,L) int(f*cos(k*pi*x/L)/L, x, -L, L);

% Define k-th sine coefficient
b = @(f,x,k,L) int(f*sin(k*pi*x/L)/L, x, -L, L);

% n-th partial sum of Fourier series
fs = @(f,x,n,L) a(f,x,0,L)/2 + ...
    symsum(a(f,x,k,L)*cos(k*pi*x/L) + b(f,x,k,L)*sin(k*pi*x/L),k,1,n);

% Fourier series absolute value |x|, 9th partial sum

f = x
pretty(fs(f,x,10,1))

% Calculate general Fourier coefficients
% k-th Fourier cosine coefficient

[A,how] = simple(a(f,x,k,1)); A
[B,how] = simple(b(f,x,k,1)); B
pretty(fs(f,x,10,1))

% Plot partial sums n=2,5,10,20,50

ezplot(fs(f,x,2,1),-1,1)
hold on
ezplot(f,-1,1)
hold off
title('Partial sum with n=2')

ezplot(fs(f,x,5,1),-1,1)
hold on
ezplot(f,-1,1)
hold off
title('Partial sum with n=5')
```

```
ezplot(fs(f,x,10,1),-1,1)
hold on
ezplot(f,-1,1)
hold off
title('Partial sum with n=10')

ezplot(fs(f,x,20,1),-1,1)
hold on
ezplot(f,-1,1)
hold off
title('Partial sum with n=20')

ezplot(fs(f,x,50,1),-1,1)
hold on
ezplot(f,-1,1)
hold off
title('Partial sum with n=50')

% Using plot for improved graph

g = inline(vectorize(fs(f,x,50,1)));
X= -1:0.001:1;
plot(X,g(X),'r')
hold on
ezplot(f,-1,1)
hold off
title('f(x)=x Partial Sum n=50')

g = inline(vectorize(fs(f,x,100,1)));
X= -1:0.0001:1;
plot(X,g(X),'r')
hold on
ezplot(f,-1,1)
hold off
title('f(x)=x Partial Sum n=100')

% Notice overshoots at end of intervals

ezplot(fs(f,x,20,1),-2,2)
hold on
ezplot(f,-2,2)
hold off
```

```
% Use Heaviside function to extend f(x)=x periodic over [-3,3]

f = heaviside(x+3)*heaviside(-1-x)*(x+2) + ...
    heaviside(x+1)*heaviside(1-x)*x + ...
    heaviside(x-1)*heaviside(3-x)*(x-2);

% Plot function and Fourier series

ezplot(fs(x,x,20,1),-3,3)
hold on
ezplot(f,-3,3)
hold off
title('Periodic extension of x and partial sum with n=20')

% Plot extended function & partial sums

X = -3:0.001:3;
g = inline(vectorize(fs(f,x,20,1)));
ezplot(f,-3,3)
hold on
plot(X,g(X),'r')
hold off
title('Periodic Extension of x and Partial Sums n=20')

% The overshoots at the discontinuities of the periodic extensions of f
% are called the Gibbs Phenomenon

% Thanks to Professor Rudolph Lorentz
```

Chapter 10

Monte Carlo Methods and Simulation

10.1 Random Numbers

10.1.1. $\ell_0 = 123456$

$x_1 = .96621\ 2243;$ $x_2 = .12917\ 3003;$ $x_3 = .01065\ 6910$

10.1 Computer Exercises

10.1.C8. 32.5%

10.1.C11. Sequence not periodic.

10.1.C13.

0	1	2	3	4	5	6	7	8	9
97	93	97	107	90	115	88	101	113	99

10.1.C15. 5.6%

10.1.C16. 200

10.2 Estimation of Areas and Volumes by Monte Carlo Techniques

10.2.1. We have $\pi \approx 4n/m$ and $|\pi - 4n/m| < 1/\sqrt{m}$. We want error $< .0005$. So let $1/\sqrt{m} < 5 \times 10^{-4}$ then $\sqrt{m} > \frac{1}{5} \times 10^4$ and $m > \frac{1}{25} \times 10^8 = 4 \times 10^6 = 4$ million

10.2 Computer Exercises

10.2.C2. 1.71828

10.2.C4. Using even function property of the integrand, exact answer is 8 since

$$
\begin{aligned}
8\int_0^1\int_0^1\int_0^1 (x^2+y^2+x^2)\,dx\,dy\,dz &= 8\int_0^1\int_0^1 \left(\frac{1}{3}+y^2+z^2\right)dy\,dz \\
&= 8\int_0^1 \left(\frac{1}{3}+\frac{1}{3}+z^2\right)dz \\
&= 8\left(\frac{1}{3}+\frac{1}{3}+\frac{1}{3}\right) = 8
\end{aligned}
$$

10.2.C6. 49.9

10.2.C7. 0.518

10.2.C9. 1.11

10.2.C10. 2.000346869

10.2.C14. 0.635

10.2.C17. **b.** 8.3

10.3 Simulation: Computer Exercises

10.3.C1. 2/3

10.3.C2. 0.898

10.3.C4. 7/16

10.3.C6. 1.05

10.3.C7. 5.24

10.3.C9. 0.996

10.3.C12. 0.6394

10.3.C14. 11.6

10.3.C15. 0.14758

10.3.C17. 0.009

10.3.C21. 24.2

10.3.C23. 0.6617

Chapter 11

Boundary-Value Problems

11.1 Shooting Method

11.1.2. We have

$$\begin{cases} x' = c_1e^t - c_2e^{-t}, x'' = c_1e^t + c_2e^{-t} = x \\ 1 = x(0) = c_1 + c_2, 2 = x(1) = c_1e + c_2e^{-1} \end{cases}$$

Hence, we obtain

$$\begin{cases} c_1 = (1-2e)/(1-e^2) \\ c_2 = (2e-e^2)/(1-e^2) \end{cases}$$

11.1.3. **a.** The general solution is $x(t) = c_1e^t + c_2e^{-t}$.
Now $0 = x(0) = c_1 + c_2$, $1 = c_1e^{\pi} + c_2e^{-\pi}$.
Hence, we obtain $c_1 = e^{\pi}/(e^{2\pi} - 1) = -c_2$, and
$x(t) = (e^{\pi+t} - e^{\pi-t})/(e^{2\pi} - 1)$.

b. The general solution is $x(t) = t^4/12 + c_1t + c_2$.
Now $1 = x(0) = c_2$, $-1 = x(1) = 13/12 + c_1$.
Hence, we obtain $x(t) = (t^4 - 25t + 12)/12$.

11.1.4. **a.** The general solution is $x(t) = c_1 \sin t + c_2 \cos t$.
Now $\alpha = x(0) = c_2$ and $\beta = x(\pi/2) = c_1$.
The solution is $x(t) = \beta \sin t + \alpha \cos t$ for all pairs (α, β).

b. Now $\alpha = x(0) = c_2$ and $\beta = x(\pi) = -c_2$.
But $c_2 = \alpha$ and $c_2 = -\beta$ only if $\alpha = -\beta$.
The solution is $x(t) = c_1 \sin t + \alpha \cos t$ where c_1 is arbitrary for all pairs (α, β) such that $\alpha + \beta = 0$.

11.1.6. The general solution is $x(t) = c_1 \sin t + c_2 \cos t$ and $x'(t) = c_1 \cos t - c_2 \sin t$.
Now $0 = x(0) = c_2$ and $t = x'(0) = c_1$.
Hence, we obtain $x(t) = z \sin t$ and $x\left(\frac{\pi}{2}\right) = z$.
Thus, we have $\varphi(t) = z$.

11.1.7. The general solution is $x(t) = c_1 \sin t + c_2 \cos t$.

Now $1 = x(0) = c_2$.

Moreover, we have $x'(t) = c_1 \cos t - \sin t$.

Let $z = x'(0) = c_1$.

Hence, the solution is $x(t) = z \sin t + \cos t$ and $\varphi(z) = x(\pi/2) = z$.

Thus, we have $\varphi(z) = z$.

11.1.8. The general solution is $x(t) = c_1(c_2 + t)^{\frac{1}{2}}$.

Now $3 = x(1) = c_1(c_2 + 1)^{\frac{1}{2}}$ and $c_1 = 3(c_2 + 1)^{-\frac{1}{2}}$.

Hence, we obtain $x(t) = 3(c_2 + t)^{\frac{1}{2}}/(c_2 + 1)^{\frac{1}{2}}$.

Let $z = x'(1) = \frac{3}{2}(c_2 + 1)^{-1}$ and $c_2 = \frac{3}{2z} - 1$.

Therefore, $x(t) = 3((\frac{3}{2z}) - 1 + t)^{\frac{1}{2}}/(\frac{3}{2z})^{\frac{1}{2}}$ and $\varphi(z) = x(2) = \sqrt{9 + 6z}$.

From $\varphi(z) = 5$ we get $z = \frac{8}{3}$ and $x = \sqrt{16t - 7}$.

11.1.9. The general solution is $x(t) = c_1 e^t + c_2 e^{-t}$.

Now $e = x(-1) = c_1 e^{-1} + c_2 e$ or $c_1 = e^2(1 - c_2)$.

Hence, we obtain $x(t) = e^2(1 - c_2)e^t + c_2 e^{-t}$.

Moreover, we have

$x'(t) = e^2(1 - c_2)e^t - c_2 e^{-t}$.

Let $z = x'(-1) = e^2(1 - c_2)e^{-1} - c_2 e = (1 - 2c_2)e$ so that $c_2 = (e - z)/(2e)$.

Therefore, $\varphi(z) = x(1) = (e^5 + e - z + e^4 z)/(2e^2)$.

11.1.10. Two ways: Use $x''(a) = z$ or $x'(b) = z$, $x''(b) = w$.

11.1.11. We obtain

$$\begin{aligned} x'' &= -e^t - \frac{4}{2}(t+1)^{-2} + c_1 \\ x' &= -e^t + 2(t+1)^{-1} + c_1 t + c_2 \\ x &= -e^t + 2\ln(t+1) + \frac{c_1}{2}t^2 + c_2 t + c_3 \end{aligned}$$

Now $-1 = x(0) = -1 + c_3 \Rightarrow c_3 = 0$

$$\begin{aligned} 3 - e + 2\ln 2 &= x(1) = -e + 2\ln 2 + \frac{c_1}{2} + c_2 &&\Rightarrow c_1 + 2c_2 = 6 \\ 6 - e^2 + 2\ln 3 &= x(2) = -e^2 + 2\ln 3 + 2c_1 + 2c_2 &&\Rightarrow c_1 + c_2 = 3 \end{aligned}$$

Clearly, $c_2 = 3$, $c_1 = 0$.

Hence, the solution is $x(t) = -e^t + 2\ln(t+1) + 3t$.

11.1.14. **a.** This is a linear problem. So two initial-value problems can be solved as in the text to obtain the solution.

The two sets of initial values would be

$$\begin{cases} x(0) = 0 \\ x'(0) = 1 \end{cases} \qquad \begin{cases} x(0) = 1 \\ x'(0) = 0 \end{cases}$$

11.1.15. Solution of $x'' = -x, x(0) = 1, x'(0) = z$ is $x(t) = \cos t + z \sin t$.

So $\varphi(z) = x(\pi) = -1$.

Since φ is constant, we can not get $\varphi(z) = 3$ by *any* choice of z!

11.1 Computer Exercises

11.1.C1. Let $c_2 = 0$ and $c_1 = 8\pi^2$.

Then $\alpha = \ln 8\pi^2$ and $\beta = \ln 8\pi^2$.

Moreover, $x'(t) = 2(c_1/2)^{\frac{1}{2}} \tan(t(c_1/2)^{\frac{1}{2}} + c_2)$ so that $x'(0) = 4\pi \tan 0 = 0$

The general solution is $x(t) = \ln 8\pi^2 - 2\ln\cos(2\pi t)$.

We solve

$$\begin{cases} x'' = e^x \\ x(0) = \ln 8\pi^2 \\ x'(0) = z \end{cases}$$

with $z_1 = -25/2$ and $z_2 = -23/2$.

11.2 A Discretization Method

11.2.1. For form (4), we have $u(t) = t$, $v(t) = 2$, $w(t) = -1$.

Hence, the typical equation by (5) would be

$$-\left(1 - \frac{h}{2}\right)x_{i-1} + 2(1 + h^2)x_i - \left(1 + \frac{h}{2}\right)x_{i+1} = -h^2 t_i$$

11.2.2. $h = 1/4 \Rightarrow n = 3$.

The tridiagonal system is

$$\begin{cases} (2 - h^2)x_1 & -x_2 & & = 0 \\ & -x_1 + (2 - h^2)x_2 & -x_3 & = 0 \\ & & -x_2 + (2 - h^2)x_3 & = 1 \end{cases}$$

$$\begin{cases} \frac{31}{16}x_1 & -x_2 & & = 0 \\ -x_1 + \frac{31}{16}x_2 & -x_3 & = 0 \\ & -x_2 + \frac{31}{16}x_3 & = 1 \end{cases}$$

Hence, we obtain $x_1 = (2 - h^2)^{-1}x_2$, $x_3 = (2 - h^2)^{-1}(1 + x_2)$ by the first and last equation.

By the second one, we have

$$[-(2 - h^2)^{-1} + (2 - h^2) - (2 - h^2)^{-1}]x_2 = (2 - h^2)^{-1} \text{ or } x_2 = 1/[-2 + (2 - h^2)^2]$$

With $h = \frac{1}{4}$, $x_2 = 256/449 \approx 0.57016$, $x_1 = 4096/13919 \approx 0.29427$, $x_3 = 11280/13919 \approx 0.81040$. Approximation more accurate near boundary since large h was used.

11.2.4. $x'' = x^2 - t + tx$ is not linear equation and so shooting will have to be iterative. *Linear Interpolation*: 3 is 2/3rds of the way between 2 and 3.5.

So we try an initial value of $x'(0)$ 2/3rds of the way between $u'(0)$ and $v'(0)$, i.e., try $x'(0) = 5/3$.

11.2.8. $(x_{i-1} - 2x_i + x_{i+1})/h^2 = (1 + t_i)x_i$, $-x_{i-1} + [2 + (1 + t_i)h^2]x_i - x_{i+1} = 0$

Hence, we obtain

$$\begin{cases} & [2 + (1 + t_1)h^2]x_1 & - x_2 & = & 0 \\ -x_{i-1} & + [2 + (1 + t_i)h^2]x_i & - x_{i+1} & = & 0 \ (2 \le i \le n-1) \\ -x_{n-1} & + [2 + (1 + t_n)h^2]x_n & & = & -1 \end{cases}$$

11.2.9. To solve

$$\begin{cases} x'' = e^t x + t^2 x' \\ x(1) = 0 \quad x(2) = 7 \end{cases}$$

Suppose

$$\begin{cases} u'' = e^t u + t^2 u' \\ u(1) = 0 \quad u'(1) = 1 \end{cases}$$

then x will be a multiple of u.

If $x = \alpha u$, we determine α from the condition $7 = x(2) = \alpha u(2)$.

The fact that $u(2) \neq 0$ follows from the increasing behavior of $u(t)$ on the interval $[1, 2]$.

11.2.11. $x_1'' = -x_1, x_1(0) = 3, x_1'(0) = z_1$ implies
$x = A\cos t + B\sin t, 3 = x(0) = A, x' = -A\sin t + B\cos t.$

Let $z_1 = x'(0) = B$.

So $x_1 = 3\cos t + z_1 \sin t,\ x_2 = 3\cos t + z_2 \sin t,$

By Eqn (10), $x = \lambda x_1 + (1-\lambda)x_2$ and $\lambda = [\beta - x_2(b)]/[x_1(b) - x_2(b)]$
$= [7-(-3)]/[(-3)-(-3)] = 10/0!!$

11.2 Computer Exercises

11.2.C1.

t	x
0.00000	1.00000 00000 00
0.11111	1.04596 43625 148
0.22222	1.08421 37270 667
0.33333	1.11366 39070 190
0.44444	1.13294 09456 874
0.55556	1.14031 38573 989
0.66667	1.13362 66672 380
0.77778	1.11023 44247 917
0.88889	1.06694 88627 917
1.00000	1.00000 00000 000

11.2.C2. **a.** Solution: $x = 1/(1+t)$

b. Solution: $x = -\log(1+t)$

Chapter 12

Partial Differential Equations

12.1 Parabolic Problems

12.1.1. **a.** Elliptic **c.** Parabolic **f.** Hyperbolic

12.1.2. Let $x = r\cos\theta$, $y = r\sin\theta$. Clearly, $\dfrac{\partial x}{\partial r} = \cos\theta$, $\dfrac{\partial y}{\partial r} = \sin\theta$, $\dfrac{\partial x}{\partial \theta} = -r\sin\theta = -y$, $\dfrac{\partial y}{\partial \theta} = r\cos\theta = x$. Now we have

$$\begin{aligned}
\frac{\partial u}{\partial \theta} &= \frac{\partial u}{\partial x}\frac{\partial x}{\partial \theta} + \frac{\partial u}{\partial y}\frac{\partial y}{\partial \theta} = -y\frac{\partial u}{\partial x} + x\frac{\partial u}{\partial y} \\
\frac{\partial^2 u}{\partial \theta^2} &= -\frac{\partial y}{\partial \theta}\frac{\partial u}{\partial x} - y\frac{\partial}{\partial \theta}\left(\frac{\partial u}{\partial x}\right) + \frac{\partial x}{\partial \theta}\frac{\partial u}{\partial y} + x\frac{\partial}{\partial \theta}\left(\frac{\partial u}{\partial y}\right) \\
&= -x\frac{\partial u}{\partial x} - y\frac{\partial^2 u}{\partial x^2}\frac{\partial x}{\partial \theta} - y\frac{\partial u}{\partial y} + x\frac{\partial^2 y}{\partial y^2}\frac{\partial y}{\partial \theta} \\
&= -x\frac{\partial u}{\partial x} - y\frac{\partial u}{\partial y} + y^2\frac{\partial^2 u}{\partial x^2} + x^2\frac{\partial^2 u}{\partial y^2} \qquad \text{(A)}
\end{aligned}$$

Similarly, we have

$$\frac{\partial u}{\partial r} = \frac{\partial u}{\partial x}\frac{\partial x}{\partial r} + \frac{\partial u}{\partial y}\frac{\partial y}{\partial r} = \cos\theta\frac{\partial u}{\partial x} + \sin\theta\frac{\partial u}{\partial y} \text{ and } r\frac{\partial u}{\partial r} = x\frac{\partial u}{\partial x} + y\frac{\partial u}{\partial y}$$

Continuing, we have

$$\begin{aligned}
\frac{\partial}{\partial r}\left(r\frac{\partial u}{\partial r}\right) &= \frac{\partial x}{\partial r}\frac{\partial u}{\partial x} + x\frac{\partial}{\partial r}\left(\frac{\partial u}{\partial x}\right) + \frac{\partial y}{\partial r}\frac{\partial u}{\partial y} + y\frac{\partial}{\partial r}\left(\frac{\partial u}{\partial y}\right) \\
&= \cos\theta\frac{\partial u}{\partial x} + \sin\theta\frac{\partial u}{\partial y} + x\frac{\partial^2 u}{\partial x^2}\frac{\partial x}{\partial r} + y\frac{\partial^2 u}{\partial y^2}\frac{\partial y}{\partial r} \\
&= x\frac{\partial u}{\partial x} + y\frac{\partial u}{\partial y} + x^2\frac{\partial^2 u}{\partial x^2} + y^2\frac{\partial^2 u}{\partial y^2} \qquad \text{(B)}
\end{aligned}$$

Adding (A) and (B), we have $r\dfrac{\partial}{\partial r}\left(r\dfrac{\partial u}{\partial r}\right) + \dfrac{\partial^2 u}{\partial \theta^2} = (x^2+y^2)\left(\dfrac{\partial^2 u}{\partial x^2} + \dfrac{\partial^2 u}{\partial y^2}\right)$.

Hence, we obtain

$$\frac{\partial^2 u}{\partial x^2} + \frac{\partial^2 u}{\partial y^2} = \frac{1}{r}\frac{\partial}{\partial r}\left(r\frac{\partial u}{\partial r}\right) + \frac{1}{r^2}\frac{\partial^2 u}{\partial \theta^2}$$

12.1.4. Equation (3) shows that $u(x, t+k)$ is a convex combination of values of $u(x,t)$ in the interval $[0,1]$. So it remains in the interval.

12.1.5. Let $u(x,t) = a^t \sin \pi x$. By (3), we have $a^{t+k} \sin x = sa^t \sin \pi(x+h) + (1-2s)a^t \sin \pi x + sa^t \sin \pi(x-h) = a^t s[\sin \pi(x+h) + \sin \pi(x-h)] + a^t[(1-2s)\sin \pi x]$ and $a^k \sin x = 2s \sin \pi x \cos \pi h + (1-2s)\sin \pi x$.

Hence, we obtain $a = [1 + 2kh^{-2}(\cos \pi h - 1)]^{1/k}$.

12.1.6. The right-hand side is changed by $b_1 + c_0$ in place of b_1 and $b_{n-1} + c_n$ replacing b_{n-1} for both (5) and (7).

12.1.7. In (6), b_1 is replaced by $b_1 + g(t)$, b_{n-1} by $b_{n-1} + g(t)$. At the level zero, $b_i = f(ih)$ for $1 \le i \le n-1$.

12.1.8. We have
$[u(x+h,t) - 2u(x,t) + u(x-h,t)]/h^2 = [u(x,t+k) - u(x,t)]/k + [u(x+h,t) - u(x,t)]/h$.
Hence, we obtain

$$u(x,t+k) = \frac{k}{h^2}(1-h)u(x+h,t) + \frac{k}{h^2}\left(\frac{h^2}{k} + h - 2\right)u(x,t) + \frac{k}{h^2}u(x-h,t).$$

12.1.9. $(u_{i+1} - u_{i-1})/(2h) + \left(\frac{\partial u}{\partial t}\right)_i = 0$ or fixed t and $1 \le i \le n-1$. Let $u_i' = \left(\frac{\partial u}{\partial t}\right)_i$.

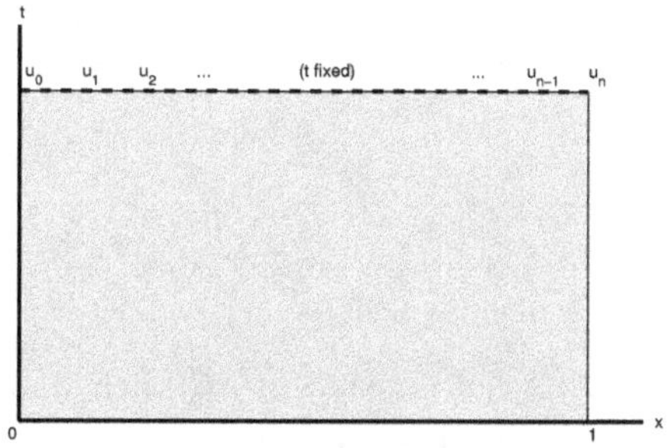

We have

$$\begin{cases} u_1' + \dfrac{1}{2h}(-u_0 + u_2) = 0 \\ u_2' + \dfrac{1}{2h}(-u_1 + u_3) = 0 \\ \vdots \\ u_{n-1}' + \dfrac{1}{2h}(-u_{n-2} + u_n) = 0 \\ u_n' + \dfrac{1}{h}(-u_{n-1} + u_n) = 0 \end{cases}$$

Then $\dfrac{dy}{dx} + \dfrac{1}{2h}\mathbf{A}\mathbf{y} = 0$ where

$$\mathbf{A} - \begin{bmatrix} 0 & 1 & & & \\ -1 & 0 & 1 & & \\ & \ddots & \ddots & \ddots & \\ & & -1 & 0 & 1 \\ & & & -2 & 2 \end{bmatrix} \quad \text{since } u_0 = 0$$

12.2 Hyperbolic Problems

12.2.1. Extend f as in textbook. Then
$u(3.3, 4) = \frac{1}{2}f(3.3 + 4) + \frac{1}{2}f(3.3 - 4) = \frac{1}{2}f(7.3) + \frac{1}{2}f(-0.7)$
$= \frac{1}{2}[f(-0.7) + f(-0.7)] = -\frac{1}{2}[f(0.7) + f(0.7)] = -f(0.7) = -(0.7)(0.3) = -0.21.$

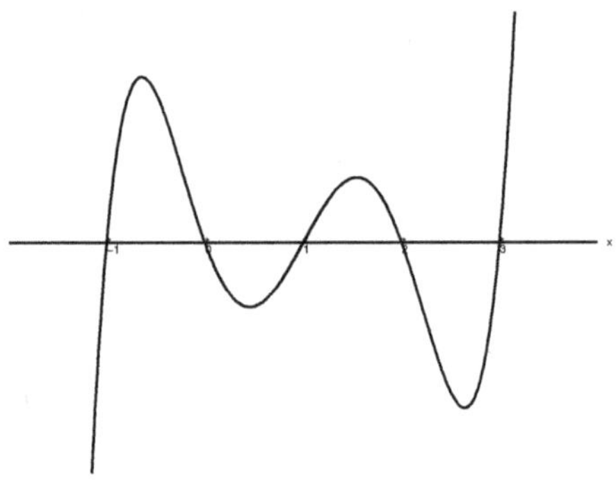

12.2.2. $u_{xx} = f''(x + at) + g''(x - at),\ u_{tt} = a^2 f''(x + at) + a^2 g''(x - at) = a^2 u_{xx}$

12.2.3. Let $u(x, t) = f(x + t) + g(x - t)$ with f and g to be determined.
Eqn. satisfied by Exercise 12.1.2.
By boundary conditions, $F(x) = f(x) + g(x)$, $G(x) = f'(x) - g'(x)$ $0 = f(t) + g(-t)$, $0 = f(1 + t) + g(1 - t)$.
So $g(t) = -f(-t)$, $g'(t) = f'(-t)$.
Hence, we obtain $F(x) = f(x) - f(-x)$, $F'(x) = f'(x) + f'(-x)$, $G(x) = f'(x) - f'(-x)$.
Adding $f'(x) = \frac{1}{2}[G(x) + F'(x)]$ and $f(x) = \frac{1}{2}\overline{G}(x) + F(x)]$ where $\overline{G}$ antiderivative of G.
Now solution is

$$u(x, t) = \frac{1}{2}[F(x + t) - F(-x + t)] + \frac{1}{2}[\overline{G}(x + t) - \overline{G}(-x + t)]$$

12.2 Computer Exercises

12.2.C1.

real function *fbar*(x)
real x, *xbar*
xbar $\leftarrow x + 2\, real(integer(-\frac{1}{2}(1+x)))$
if *xbar* < 0 **then**
  fbar $\leftarrow -f(-xbar)$
else
  fbar $\leftarrow f(xbar)$
end if
end function *fbar*

12.3 Elliptic Problems

12.3.5. $20(u_{i+1,j} - 2u_{ij} + u_{i-1,j})/h^2 - 30(u_{i,j+1} - 2u_{ij} + u_{i,j-1})/h^2$
$+[5/(x_i + y_j)](u_{i+1,j} - u_{i-1,j})/(2h) + (1/y_j)\ (u_{i,j+1} - u_{i,j-1})/(2h) = 69.$
Hence, we obtain

$$\left(20 + \frac{2.5h}{x_i + y_j}\right) u_{i+1,j} + \left(20 - \frac{2.5h}{x_i + y_j}\right) u_{i-1,j}$$
$$+\left(-30 + \frac{0.5h}{y_j}\right) u_{i,j+1} + \left(-30 + \frac{0.5h}{y_j}\right) u_{i,j-1} + 20u_{ij} = 69h^2$$

$$u(0, \tfrac{1}{2}) = (-2 + \sqrt{3})/30 \approx -8.932 \times 10^{-3}$$
$$u(\tfrac{1}{2}, \tfrac{1}{2}) = (7 + 4\sqrt{3})/30 \approx 4.643 \times 10^{-1}$$

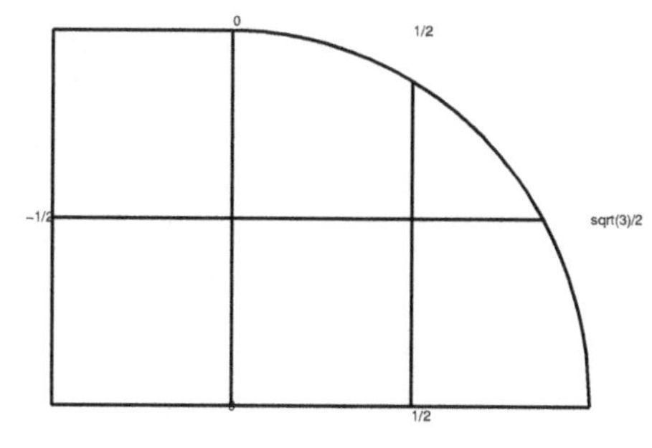

12.3.6.

$$\begin{cases} -4u_1 + u_3 + u_2 = 2 \\ u_1 - 4u_2 + u_4 = 5 - \dfrac{4}{3} \\ u_1 + u_4 - u_3 = 5 + \dfrac{4}{3} \\ u_2 + u_3 - 4u_4 = 8 - \dfrac{5}{3} \end{cases}$$

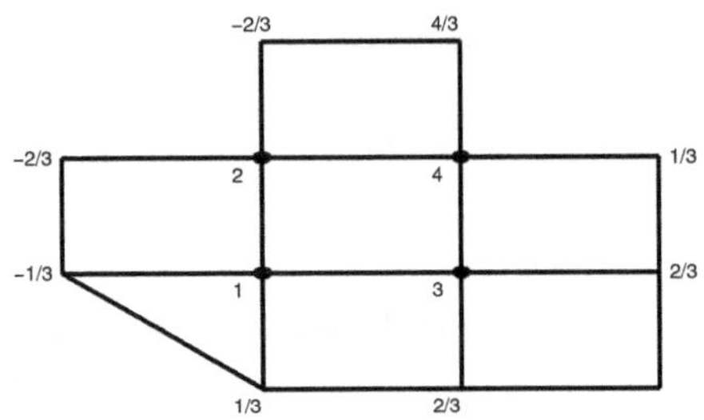

$$\mathbf{A} = \begin{bmatrix} -4 & 1 & 1 & 0 \\ 1 & -4 & 0 & 1 \\ 1 & 0 & -4 & 1 \\ 0 & 1 & 1 & -4 \end{bmatrix}$$

$$\mathbf{b} = \begin{bmatrix} 2 \\ \dfrac{11}{3} \\ \dfrac{19}{3} \\ \dfrac{19}{3} \end{bmatrix}, \quad \mathbf{u} = \begin{bmatrix} u_1 \\ u_2 \\ u_3 \\ u_4 \end{bmatrix}$$

12.3 Computer Exercises

12.3.C5.

$$\begin{bmatrix} 3 & -1 & 0 & 0 & 0 & -1 & 0 & 0 & 0 & 0 \\ -1 & 4 & -1 & 0 & 0 & 0 & -1 & 0 & 0 & 0 \\ 0 & -1 & 4 & -1 & 0 & 0 & 0 & -1 & 0 & 0 \\ 0 & 0 & -1 & 4 & -1 & 0 & 0 & 0 & 0 & 0 \\ 0 & 0 & 0 & -1 & 3 & 0 & 0 & 0 & 0 & 0 \\ -1 & 0 & 0 & 0 & 0 & 3 & -1 & 0 & -1 & 0 \\ 0 & -1 & 0 & 0 & 0 & -1 & 4 & -1 & 0 & -1 \\ 0 & 0 & -1 & 0 & 0 & 0 & -1 & 4 & 0 & 0 \\ 0 & 0 & 0 & 0 & 0 & -1 & 0 & 0 & 3 & -1 \\ 0 & 0 & 0 & 0 & 0 & 0 & -1 & 0 & -1 & 4 \end{bmatrix} \begin{bmatrix} u_1 \\ u_2 \\ u_3 \\ u_4 \\ u_5 \\ u_6 \\ u_7 \\ u_8 \\ u_9 \\ u_{10} \end{bmatrix} = \begin{bmatrix} 100 \\ 100 \\ 100 \\ 100 \\ 100 \\ 0 \\ 0 \\ 0 \\ 0 \\ 0 \end{bmatrix}$$

by Gauss-Seidel, we obtain

$$\begin{matrix} 18.41^\circ & 13.75^\circ & & & \\ 41.47^\circ & 36.60^\circ & 24.41^\circ & & \\ 69.41^\circ & 66.77^\circ & 61.05^\circ & 53.01^\circ & 51.00^\circ \end{matrix}$$

Chapter 13

Minimization of Functions

13.1 One Variable Case

13.1.1. $\dfrac{\partial F}{\partial x_1} = 2x_1 - 4 = 0,\ \dfrac{\partial F}{\partial x_2} = 6x_2 - 6 = 0,\ \dfrac{\partial F}{\partial x_3} = 4x_3 + 8 = 0,$
$F(2, 1, -2) = -15$; $F = (x_1 - 2)^2 + 3(x_2 - 1)^2 + 2(x_3 + 2)^2 - 15$, $F(0, 0, -2) = -8$;
$F(2, 0, -2) = -12$.

13.1.2. Let $x = u + v$, $y = u - v$.

Then $F = 16u^2 + 36v^2 - 36u$, $\dfrac{\partial F}{\partial u} = 32u - 36 = 0$, $\dfrac{\partial F}{\partial v} = 72v = 0$.

Thus, we have $F(\frac{9}{8}, \frac{9}{8}) = -20.25$.

13.1.4. **Case** $n = 2$. Let $x* \in [a, b']$ where $b' = \frac{1}{2}(a + b) + 2\delta$.

Now $\hat{x} = (a + b')/2 = (3a + b)/4 + \delta$.

Similarly, if $x* \in [a', b]$ where $a' = \frac{1}{2}(a + b) - 2\delta$, then $\hat{x} = (a' + b)/2 = (a + 3b)/4 - \delta$.

In each case, half of the width of the final interval is $\frac{1}{4}(b - a) \pm \delta$.

Case $n = 3$. Let $x* \in [a, b']$ where $b' = a + \frac{2}{3}(b - a)$.

Suppose we always take the left subinterval.

The interval $[a, b]$ is now $[a, a + \frac{2}{3}(b - a)]$ and $x* \in [a, b']$ where $b' = a' + 2\delta$,
$a' = a + \frac{1}{3}(b - a)$.

Hence, the final interval is $[a, a + \frac{1}{3}(b - a) + 2\delta]$.

Now $\hat{x}$ is the midpoint $(5a + b)/6 + \delta$.

Similarly, if we always take the right subinterval $\hat{x} = (a + 5b)/6 + \delta$.

If $x*$ is in first the left and then the middle subinterval, the final interval is
$[a + \frac{1}{3}(b - a) + 2\delta,\ a + \frac{2}{3}(b - a)]$ and $\hat{x} = (b + a)/2 + \delta$.

If $x*$ is in first the right and then the middle subinterval, $\hat{x} = (b + a)/2 - \delta$.

Clearly, half the width of the final interval is $\frac{1}{6}(b - a) \pm \delta$ in all cases.

13.1.5. **a.** Initially, $[a,b]=[0,10]$. for $n=4$, $\Delta=\dfrac{\lambda_2}{\lambda_4}(b-a)=4$, $a'=a+\Delta=4$, $b'=b-\Delta=6$, $F(a')=-6\le F(b')=2$, and set $[a,b]=[0,6]$.
For $n=3$, $\Delta=\dfrac{\lambda_1}{\lambda_3}(b-a)=2$, $a'=a+\Delta=2$, $b'=b-\Delta=4$, $F(a')=-6\ge F(b')=-6$, and set $[a,b]=[2,6]$.
For $n=2$, $F(a)=-6\le F(b)=2$ and set $b=\frac{1}{2}(a+b)+2\delta=4.5$.
We have the final interval $[2,4.5]$ and set $\hat{x}=\frac{1}{2}(a+b)=3.25$.
The exact minimum is 3.

13.1.7. Letting $\lambda_n=\lambda^n$ in $\lambda_n-\lambda_{n-1}-\lambda_{n-2}=0$, we have $\lambda^{n-2}(\lambda^2-\lambda-1)=0$.
Hence, we obtain $\alpha=\frac{1}{2}(1+\sqrt{5})$ and $\beta=\frac{1}{2}(1-\sqrt{5})$ satisfy this equation.
Let $\lambda_n=A\alpha^n+B\beta^n$.
Now $\lambda_1=1$ gives $1=A+B$ and $\lambda_2=1$, $1=A\alpha+B\beta$.
The values $A=\alpha/\sqrt{5}$ and $B=-\beta/\sqrt{5}$ satisfy these two equations.
Now as $n\to\infty$,

$$\frac{\lambda_n}{\lambda_{n-1}}=\frac{A\alpha^n+B\beta^n}{A\alpha^{n-1}+B\beta^{n-1}}=\frac{A+B(\beta/\alpha)^n}{A(1/\alpha)+B(\beta/\alpha)^n(1/\beta)}\to\alpha$$

since $|\beta/\alpha|=|(1-\sqrt{5})/(1+\sqrt{5})|\approx 0.38<1$ and $(\beta/\alpha)^n\to 0$, as $n\to\infty$.
Clearly, we have

$$\lambda_n=\frac{1}{\sqrt{5}}\alpha^n-\frac{1}{\sqrt{5}}\beta^n=\frac{1}{\sqrt{5}}\left[\left(\frac{1+\sqrt{5}}{2}\right)^n-\left(\frac{1-\sqrt{5}}{2}\right)^n\right]$$

Moreover, we obtain

$$\lim_{n\to\infty}\frac{\lambda_{n+1}}{\lambda_n}=\lim_{n\to\infty}\frac{\lambda_n}{\lambda_{n-1}}=\alpha=\frac{1+\sqrt{5}}{2}$$

Thus, we have agreement with Equations (10) and (11) of Section 3.3.

13.1.9. By (6), $y+rb=a+r^2(b-a)+rb=ar+b$ since $r^2+r=1$.
Moreover, $r(y+rb)=a+r(b-a)=x$.
Thus, we have $yr+r^2b=x$ or $y+r^2(b-y)=x$.

13.1.10. Let $\ell=b-a$ and $r=\frac{1}{2}(\sqrt{5}-1)$.

We have

# evaluations	2	3	4	$\cdots$	n
width of interval	$r\ell$	$r^2\ell$	$r^3\ell$	$\cdots$	$r^{n-1}\ell$
error	$\frac{1}{2}r\ell$	$\frac{1}{2}r^2\ell$	$\frac{1}{2}r^3\ell$	$\cdots$	$\frac{1}{2}r^{n-1}\ell$

Require $\frac{1}{2}r^{n-1}\ell\le 10^{-k}$.
Hence, we obtain $\log\frac{1}{2}+(n-1)\log r+\log\ell\le -k$ or
$n-1\ge(-k-\log\ell+\log 2)/(-\log_{10}r)$ since $\log r<0$.
The number of evaluations is $n\ge 1+(k+\log\ell-\log 2)/|\log r|$.

13.1.11. $n \geq 1 + (10 + \log 1 - \log 2)/|\log 0.61803| = 47.41$. Hence, 48 evaluations are needed.

13.1.13. Minimum point of F is a root of F'. Newton's method to find root of F':

$$x_{n+1} = x_n - \frac{F'(x_n)}{F''(x_n)}$$

Formula does *not* involve F itself.

13.1.14. To find minimum of F, look for root of F'. Secant method to find root of F' is

$$x_{n+1} = x_n - F'(x_n)\left[\frac{x_n - x_{n-1}}{F'(x_n) - F'(x_{n-1})}\right]$$

Formula does *not* involve F.

13.1.15. **b.** Square both sides to obtain $r^2 = 1 + \sqrt{1 + \sqrt{1 + \ldots}} = 1 + r$.

d. $1 + r^{-1} + r^{-2} + \ldots = (1 - r^{-1})^{-1}$ by series expansion.

Hence, we obtain $r = (1 - r^{-1})^{-1} - 1 = \dfrac{1}{r-1}$ or $r^2 = r + 1$.

13.2 Multivariate Case

13.2.1. **a.** $F = x_1^2 - x_1x_2 + x_2^2 + 3x_1 + 6x_2 - 4 = (x - \frac{1}{2}y + \frac{3}{2})^2 + \frac{3}{4}(y + 5)^2$, always positive. Minimum exists. Or

$$\mathbf{G} = \begin{bmatrix} 2x_1 - x_2 + 3 \\ -x_1 + 2x_2 + 6 \end{bmatrix}, \qquad \mathbf{H} = \begin{bmatrix} 2 & -1 \\ -1 & 2 \end{bmatrix}$$

$\mathbf{x}^T\mathbf{H}\mathbf{x} = x_1^2 + x_2^2 + (x_2 - x_1)^2 > 0$ for all $x_1/0$, $x_2/0$.

$\mathbf{H}$ positive definite. Thus, a minimum exists.

b. $F = x_1^2 - 3x_1x_2 + x_2^2 + 7x_1 + 3x_2 + 5 = (x_1 - \frac{3}{2}x_2)^2 - \frac{5}{4}\, x_2^2 + 7x_1 + 3x_2 + 5$. Let $x_1 = \frac{3}{2}x_2$, $F = -\frac{5}{4}x_2^2 + \frac{27}{2}x_2 + 5 \to -\infty$ as $x_2 \to \infty$. Minimum does not exist.

13.2.2. Let $Q(\mathbf{x}) = 3x_1^2 - 2x_1x_2 + x_2^2 + 3x_1 - 4x_2 + 7$.

$$\mathbf{G} = \begin{bmatrix} 6x_1 - 2x_2 + 3 \\ [-2x_1 + 2x_2 - 4 \end{bmatrix}, \qquad \mathbf{H} = \begin{bmatrix} 6 & -2 \\ -2 & 2 \end{bmatrix}$$

Clearly, $Q(\mathbf{x}) = 7 + \mathbf{G}(0)^T\mathbf{x} + \frac{1}{2}\mathbf{x}^T\mathbf{H}(0)\mathbf{x}$ so we solve $\mathbf{H}(0)\mathbf{x} = -\mathbf{G}(0)$ or

$$\begin{bmatrix} 6 & -2 \\ -2 & 2 \end{bmatrix}\begin{bmatrix} x_1 \\ x_2 \end{bmatrix} = -\begin{bmatrix} 3 \\ -4 \end{bmatrix}$$

Hence, the local minimum point is $\left(\frac{1}{4}, \frac{9}{4}\right)$.

13.2.3.

$$\mathbf{G} = \begin{bmatrix} e^x \cos y - y/(x+1) \\ -e^x \sin y - \ln(x+1) \end{bmatrix}, \qquad \mathbf{G}(0) = \begin{bmatrix} 1 \\ 0 \end{bmatrix}$$

$$\mathbf{H} = \begin{bmatrix} e^x \cos y + y/(x+1)^2 & -e^x \sin y - 1/(x+1) \\ -e^x \sin y - 1/(x+1) & -e^x \cos y \end{bmatrix}$$

$$\mathbf{H}(0) = \begin{bmatrix} -1 & -1 \\ -1 & -1 \end{bmatrix}$$

$$\begin{aligned} F(x,y) &= 1 + (1,0)\begin{bmatrix} x \\ y \end{bmatrix} + \frac{1}{2}(x,y)\begin{bmatrix} 1 & -1 \\ -1 & -1 \end{bmatrix}\begin{bmatrix} x \\ y \end{bmatrix} + \cdots \\ &= 1 + xg - xy + \frac{1}{2}x^2 - \frac{1}{2}y^2 + \cdots \end{aligned}$$

13.2.6. The slope of the tangent is $\frac{dy}{dx} = -\frac{F_x}{F_y} \equiv m_1$. The gradient has direction numbers F_x and F_y and its slope is $\frac{F_y}{F_x} \equiv m_2$. The condition of prependicularity $m_1 m_2 = -1$ is met.

13.2.7. **b.**

$$\mathbf{G}(\mathbf{z}) = (3,0,0)^T \text{ where } \mathbf{z} = (0,1,1)^T$$

$$\mathbf{H}(\mathbf{z}) = \begin{bmatrix} 4 & 3 & 0 \\ 3 & 0 & 1 \\ 0 & 1 & -1 \end{bmatrix}$$

$$\begin{aligned} F(\mathbf{x}) &= F(\mathbf{z}) + \mathbf{G}(\mathbf{z})^T(\mathbf{x}-\mathbf{z}) + \frac{1}{2}(\mathbf{x}-\mathbf{z})^T\mathbf{H}(\mathbf{z})(\mathbf{x}-\mathbf{z}) + \cdots \\ &= \frac{3}{2} - \frac{1}{2}x_2 + 3x_1x_2 + x_2x_3 + 2x_1^2 - \frac{1}{2}x_3^2 + \cdots \end{aligned}$$

13.2.9. **a.** $\frac{\partial F}{\partial x} = 2xy - 2, \ \frac{\partial F}{\partial y} = x^2 + 1, \ \mathbf{G}(1,0) = \begin{bmatrix} -2 \\ 2 \end{bmatrix}$

b. $\frac{\partial F}{\partial x} = y + z^2 + 2xz, \ \frac{\partial F}{\partial y} = x + z^2, \ \frac{\partial F}{\partial z} = 2yz + x^2, \ \mathbf{G}(1,2,1) = \begin{bmatrix} 5 \\ 2 \\ 5 \end{bmatrix}$

13.2.10.

$$\mathbf{G} = \begin{bmatrix} 2y^2x^2 \sin x \cos x \\ 2yz^2(1+\sin^2 x) + 2(y+1)(z+3)^2 \\ 2y^2z(1+\sin^2 x) + 2(y+1)^2(z+3) \end{bmatrix}$$

13.2.12. $F(x_1, x_2) = 7 + (5,3)\begin{bmatrix} x_1 \\ x_2 \end{bmatrix} + \frac{1}{2}(x_1, x_2)\begin{bmatrix} 6 & 6 \\ 6 & -4 \end{bmatrix}\begin{bmatrix} x_1 \\ x_2 \end{bmatrix} = c + \mathbf{b}^T\mathbf{x} + \frac{1}{2}\mathbf{x}^T\mathbf{A}\mathbf{x}.$

Solve $\mathbf{A}\mathbf{x} = -\mathbf{b}$ or $\begin{bmatrix} 6 & 6 \\ 6 & -4 \end{bmatrix}\begin{bmatrix} x_1 \\ x_2 \end{bmatrix} = -\begin{bmatrix} 5 \\ 3 \end{bmatrix}$ obtaining $\left(-\frac{19}{30}, -\frac{1}{5}\right)$.

Chapter 14

Linear Programming Problems

14.1 Standard Forms and Duality

14.1.2. Maximize: $-5x_1 - 6x_2 + 2x_3$

$$\text{Constraints:} \begin{cases} -2x_1 + 3x_2 \leq -5 \\ x_1 + x_2 \leq 15 \\ 2x_1 - x_2 + x_3 \leq 25 \\ -x_1 - x_2 + x_3 \leq -1 \\ x_1 \geq 0,\ x_2 \geq 0,\ x_3 \geq 0 \end{cases}$$

14.1.4. **a.** $z(\frac{3}{2}, 0) = \frac{3}{2}$, $z(2,0) = 2$, $z(\frac{5}{2}, \frac{1}{2}) = \frac{7}{2}$, $z(0,3) = 6$ minimum value 1.5 at $(1.5, 0)$

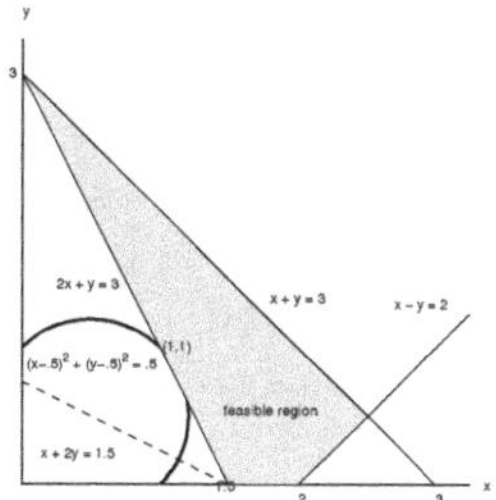

14.1.5. **b.** Maximize: $-3x + 2y - 5z$

$$\text{Constraints: } \begin{cases} -x - y - z \leq -4 \\ x - y - z \leq 2 \\ -x + y + z \leq -2 \\ x \geq 0,\ y \geq 0,\ z \geq 0 \end{cases}$$

14.1.6. **a.** Maximize: $2x_1 + 2x_2 - 6x_3 - x_4$

$$\text{Constraints: } \begin{cases} 3x_1 + x_4 = 25 \\ x_1 + x_2 + x_3 + x_4 = 20 \\ -4x_1 - 6x_3 + x_5 = -5 \\ -2x_1 - 3x_3 - 2x_4 + x_6 = 0 \\ x_1,\ x_2,\ x_3,\ x_4,\ x_5,\ x_6 \geq 0 \end{cases}$$

b. Minimize: $25y_1 + 20y_2 - 5y_3$

$$\text{Constraints: } \begin{cases} 3y_1 + y_2 - 4y_3 - 2y_3 \leq 2 \\ y_2 \leq 2 \\ y_2 - 6y_3 - 3y_4 \leq -6 \\ y_1 + y_2 - 2y_4 \leq -1 \\ y_1,\ y_2,\ y_3,\ y_4 \geq 0 \end{cases}$$

14.1.7. Maximum of 36 at $(2, 6)$.

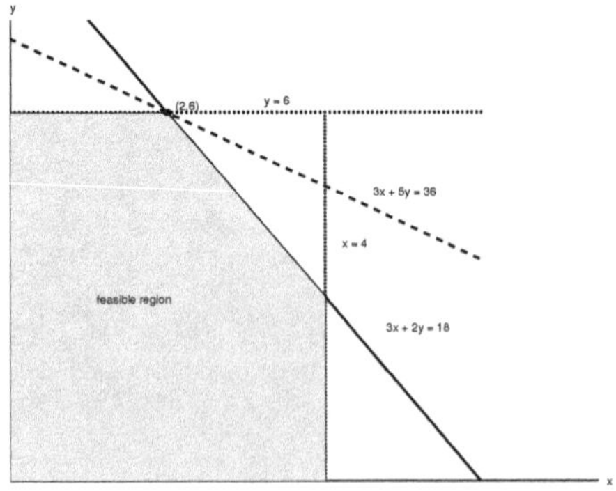

14.1.8. Dual Problem:

$$\begin{cases} \text{Minimize:} & 4y_1 + 6y_2 + 18y_3 \\ \text{Constraints:} & \begin{cases} y_1 + 3y_3 \geq & 3 \\ y_2 + 2y_3 \geq & 5 \\ y_1 \geq 0,\ y_2 \geq 0,\ y_3 \geq 0 \end{cases} \end{cases}$$

Vertices which satisfy constraints are $(3,5,0)$, $(0,3,1)$, $(0,0,5/2)$ with values 42, 36, 45, respectively. Hence, minimum of 36 at $(0,3,1)$.

14.1.11. Minimize: ε

$$\text{Constraints:} \begin{cases} x_1 - x_2 - y_1 + y_2 - \varepsilon \leq & 0 \\ -x_1 + x_2 + y_1 - y_2 - \varepsilon \leq & 0 \\ x_1 - x_2 - 3y_1 + 3y_2 \leq & 0 \\ -x_1 + x_2 + y_1 - y_2 \leq & -2 \\ x_1,\ x_2,\ y_1,\ y_2 \geq 0,\ \varepsilon \geq 0 \end{cases}$$

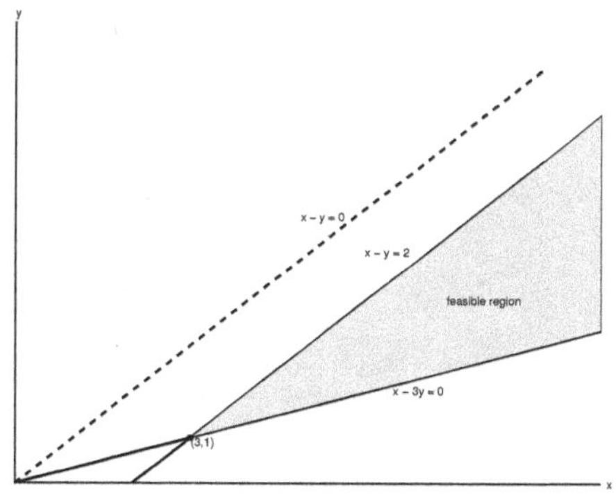

or graphical method. Minimum 2 for $(x, x-2)$ where $x \geq 3$.

14.1.13. **a.** Maximum of 18 occurs at $(9,0)$.

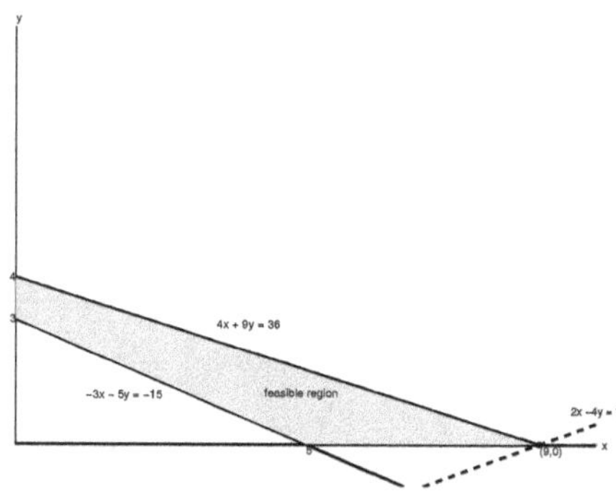

c. Unbounded solution since $3x_1 + 2x_2 \to \infty$ as x_1 and $x_2 \to -\infty$.

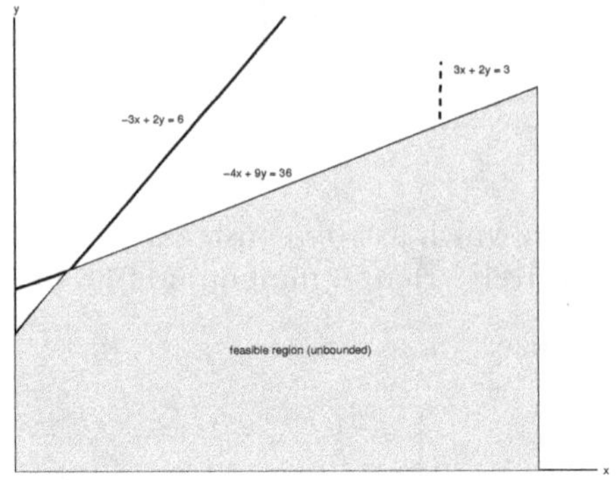

f. No solution since feasible region null for $x_1 \geq 0$ and $x_2 \geq 0$.

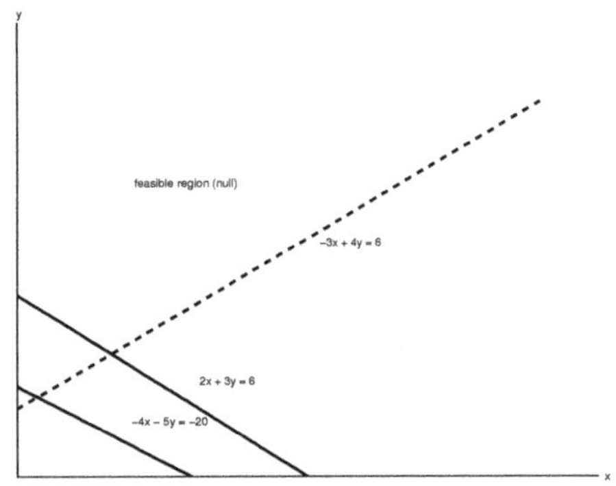

h. Maximum of $\dfrac{54}{5}$ at $(\dfrac{18}{5}, 0)$.

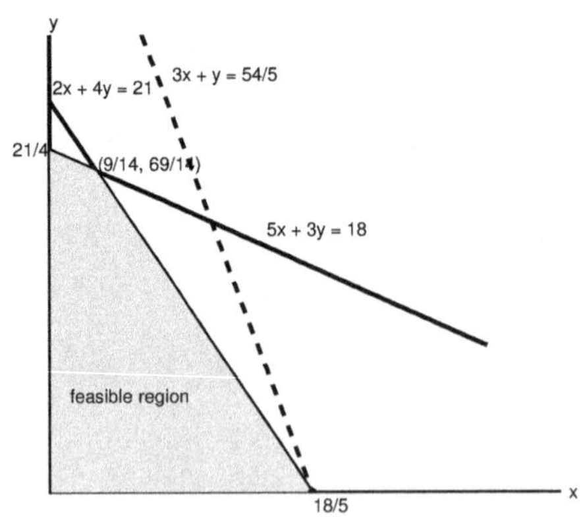

14.1.14. Maximum of 88 at $(24, 32)$ of 2-dimension problem. Maximum of 100 at $(24, 32, -124)$ of original problem.

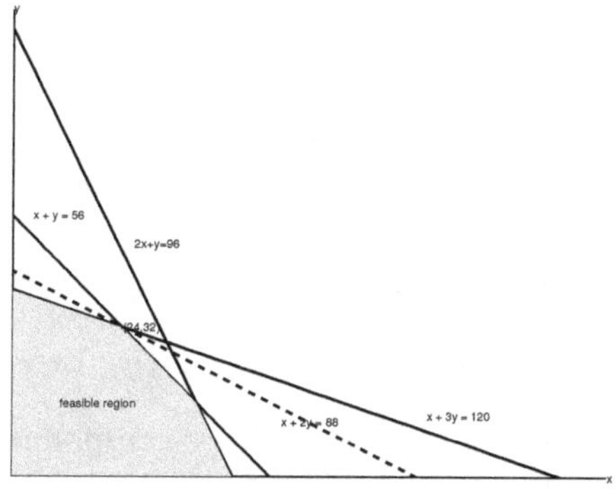

$$\begin{cases} \text{Maximize:} \quad x + 2y \\ \text{Constraints:} \begin{cases} x + 3y \leq 120 \\ x + y \leq 56 \\ 2x + y \leq 96 \end{cases} \end{cases}$$

14.1.17. Its feasible set is empty.

14.1 Computer Exercises

14.1.C1.

	Felt	Straw
Texas Hatters	0	200
Lone Star Hatters	150	0
Lariat Ranch Wear	150	0

14.1.C3. By a non-integer linear programming routine, 4.3 aircraft from α and 1.7 from β! By graphical method, 3 aircraft from β and none from α for 750 students at $13.50 each.

14.1.C5. Cost 50¢ for 1.6 ounces of food one, 1 ounce of food three, and none of food two.

14.2 Simplex Method

14.2.1. In the given problem, we do not insist that $\mathbf{x} \geq 0$. So let $x_j = y_j - y_0$ with $y_j \geq 0$ and $y_0 \geq 0$. Then $\mathbf{c}^T\mathbf{x} = \sum_{j=1}^{n} c_j x_j = \sum_{j=1}^{n} c_j(y_j - y_0) = \sum_{j=1}^{n} c_j y_j - y_0 \sum_{j=1}^{n} c_j$.

Also, $b_i \geq \sum_{j=1}^{n} a_{ij}x_j = \sum_{j=1}^{n} a_{ij}y_j - y_0 \sum_{j=1}^{n} a_{ij}$. So let $c_0 = -\sum_{j=1}^{n} c_j$ and $a_{i0} = -\sum_{j=1}^{n} a_{ij}$.

Then we want to maximize $\sum_{j=0}^{n} c_j y_j$ subject to constraints $\sum_{j=0}^{n} a_{ij}y_j \leq b_i$, $y_i \geq 0$ $(0 \leq i \leq n)$. (Since $0 \leq y_j = x_j + y_0$, we only need $y_0 \geq -x_j$ for all j.)

14.2.2. The vertices are in one-to-one correspondence with the linearly independent sets of column vectors in $\mathbf{A}$. There are at most 2^n such sets.

14.2.5. First Primal Form

$$\begin{cases} \text{Maximize:} & (-\mathbf{b}^T)\mathbf{y} \\ \text{Constraints:} & (-\mathbf{A}^T)\mathbf{y} \leq -\mathbf{c},\ \mathbf{y} \geq 0 \end{cases}$$

Second Primal Form

$$\begin{cases} \text{Maximize:} & (-\mathbf{b}^T)\mathbf{y} \\ \text{Constraints:} & [-\mathbf{A}^T \quad \mathbf{I}] \begin{bmatrix} \mathbf{y} \\ \hat{\mathbf{y}} \end{bmatrix} = \mathbf{c} \end{cases}$$

where $\hat{\mathbf{y}} = (y_{n+1}, y_{n+2}, \ldots, y_{2n})^T$

14.2.6. Given $\mathbf{A}\mathbf{x} = \mathbf{b}$. Let $y_j = x_j + y_{n+1}$. Now $\sum_{j=1}^{n} a_{ij}x_j - b_i = \sum_{j=1}^{n} a_{ij}y_j - y_{n+1} \sum_{j=1}^{n} a_{ij} - b_i$.

Minimize: y_{n+1}

$$\text{Constraints: } \begin{cases} \sum_{j=1}^{n} a_{ij}y_j + \left(-\sum_{j=1}^{n} a_{ij}\right) y_{n+1} = b_i \quad (1 \leq i \leq n+1) \\ y \geq 0 \end{cases}$$

14.2 Computer Exercises

14.2.C1. **b.** $\mathbf{x} = (0, 0, 5/3, 2/3, 0)^T$

c. $\mathbf{x} = (0, 8/3, 5/3)^T$

14.3 Inconsistent Linear Systems

14.3.1. **a.** Maximize: $-\sum_{i=1}^{4}(u_i + v_i)$

$$\text{Constraints: } \begin{cases} 5y_1 + 2y_2 - 7y_4 - u_1 + v_1 = 6 \\ y_1 + y_2 + y_3 - 3y_4 - u_2 + v_2 = 2 \\ 7y_2 - 5y_3 - 2y_4 - u_3 + v_3 = 11 \\ 6y_1 + 9y_3 - 15y_4 - u_4 + v_4 = 9 \\ u \geq 0,\ v \geq 0,\ y \geq 0 \end{cases}$$

b. Minimize: ε

$$\text{Constraints: } \begin{cases} 5y_1 + 2y_2 - 7y_4 - \varepsilon \leq 6 \\ y_1 + y_2 + y_3 - 3y_4 - \varepsilon \leq 2 \\ 7y_2 - 5y_3 - 2y_4 - \varepsilon \leq 11 \\ 6y_1 + 9y_3 - 15y_4 - \varepsilon \leq 9 \\ -5y_1 - 2y_2 + 7y_4 - \varepsilon \leq -6 \\ -y_1 - y_2 - y_3 + 3y_4 - \varepsilon \leq -2 \\ -7y_2 + 5y_3 + 2y_4 - \varepsilon \leq -11 \\ -6y_1 - 9y_3 + 15y_4 - \varepsilon \leq -9 \\ \varepsilon \geq 0,\ y \geq 0 \end{cases}$$

14.3.3. Take m points x_i $(i = 1, 2, \ldots, m)$. Let $p(x) = \sum_{j=0}^{n} a_j x^j$.

Minimize: ε

$$\text{Constraints: } \begin{cases} \sum_{j=0}^{n} a_j x_i^j \leq f(x_i) \quad (1 \leq i \leq m) \\ \sum_{j=0}^{n} a_j x_i^j + \varepsilon \geq f(x_i) \quad (1 \leq i \leq m) \\ \varepsilon \geq 0 \end{cases}$$

14.3.4. Minimize: $u_1 + v_1 + u_2 + v_2 + u_3 + v_3$

$$\text{Constraints: } \begin{cases} y_1 - y_2 - u_1 + v_1 = 4 \\ 2y_1 - 3y_2 + y_3 - u_2 + v_2 = 7 \\ y_1 + y_2 - 2y_3 - u_3 + v_3 = 2 \\ y_1,\ y_2,\ y_3 \geq 0,\ u_1,\ u_2,\ u_3 \geq 0,\ v_1,\ v_2,\ v_3 \geq 0 \end{cases}$$

14.3 Computer Exercises

14.3.C1. **a.** $y_1 = 0.353$, $y_2 = 2.118$, $y_3 = 0.765$, $y_4 = 0$, $u_1 = 0$, $u_2 = 1.235$, $u_3 = u_4 = 0$, $v_1 = v_2 = v_3 = v_4 = 0$ so that $x_1 = 0.353$, $x_2 = 2.118$, $x_3 = 0.765$.

b. $y_1 = 0.671$, $y_2 = 1.768$, $y_3 = 0.453$, $y_4 = 0$, $\varepsilon = 0.892$ so that $x_1 = 0.671$, $x_2 = 1.768$, $x_3 = 0.453$.

14.3.C3. $p(x) = 1.0001 + 0.9978x + 0.51307x^2 + 0.13592x^3 + 0.071344x^4$.

Appendix A

Upper and Lower Sums

The existence of the definite integral of a nonnegative function f on a closed interval $[a, b]$ is based on an interpretation of that integral as the area under the graph of f. The definite integral is defined by means of two concepts, the *lower sums* of f and the *upper sums* of f; these are approximations to the area under the graph.
Let P be a **partition** of the interval $[a, b]$ given by

$$P = \{a = x_0 < x_1 < x_2 < \cdots < x_{n-1} < x_n = b\}$$

with partition points $x_0, x_1, x_2, \ldots, x_n$ that divide the interval $[a, b]$ into n subintervals $[x_i, x_{i+1}]$. Now denote by m_i the **greatest lower bound** (*infimum* or inf) of $f(x)$ on the subinterval $[x_i, x_{i+1}]$. In symbols, we have

$$m_i = \inf\{f(x) : x_i \leq x \leq x_{i+1}\}$$

Likewise, we denote by M_i the **least upper bound** (*supremum* or sup) of $f(x)$ on $[x_i, x_{i+1}]$. Thus, we obtain

$$M_i = \sup\{f(x) : x_i \leq x \leq x_{i+1}\}$$

The **lower sums** and **upper sums** of f corresponding to the given partition P are defined to be

$$L(f; P) = \sum_{i=0}^{n-1} m_i(x_{i+1} - x_i)$$

$$U(f; P) = \sum_{i=0}^{n-1} M_i(x_{i+1} - x_i)$$

If f is a positive function, these two quantities can be interpreted as estimates of the area under the curve for f. These sums are shown in Figure 1.

Example. What are the numerical values of the upper and lower sums for $f(x) = x^2$ on the interval $[0, 1]$ if the partition is $P = \{0, \frac{1}{4}, \frac{1}{2}, \frac{3}{4}, 1\}$?

Solution. We want the value of

$$U(f; P) = M_0(x_1 - x_0) + M_1(x_2 - x_1) + M_2(x_3 - x_2) + M_3(x_4 - x_3)$$

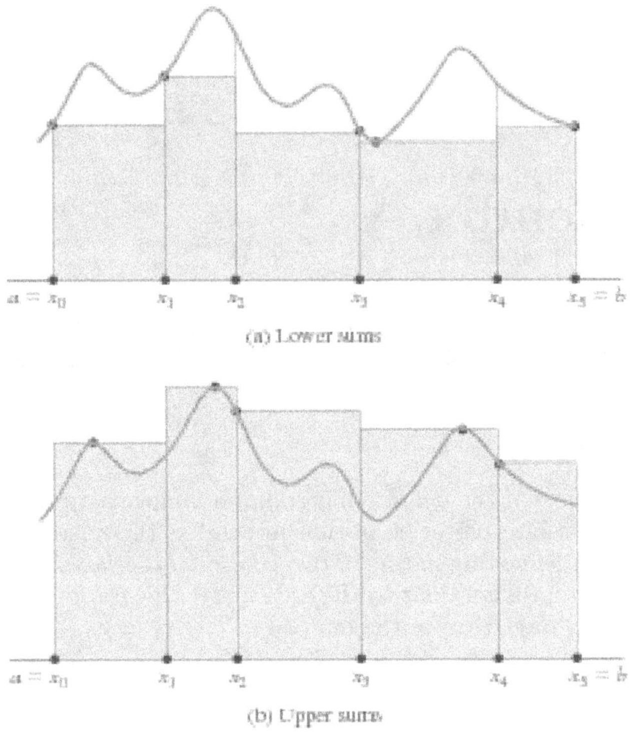

Figure A.1: Illustrating lower and upper sums

Since f is increasing on $[0,1]$, $M_0 = f(x_1) = \frac{1}{16}$. Similarly, $M_1 = f(x_2) = \frac{1}{4}$, $M_2 = f(x_3) = \frac{9}{16}$, and $M_3 = f(x_4) = 1$. The widths of the subintervals are all equal to $\frac{1}{4}$. Hence, we obtain

$$U(f;P) = \tfrac{1}{4}\left(\tfrac{1}{16} + \tfrac{1}{4} + \tfrac{9}{16} + 1\right) = \tfrac{15}{32}$$

In the same way, we find that $m_0 = f(x_0) = 0$, $m_1 = \frac{1}{16}$, $m_2 = \frac{1}{4}$, and $m_3 = \frac{9}{16}$. Hence, we have

$$L(f;P) = \tfrac{1}{4}\left(0 + \tfrac{1}{16} + \tfrac{1}{4} + \tfrac{9}{16}\right) = \tfrac{7}{32}$$

If we had no other way of calculating $\int_0^1 x^2\,dx$, we would take a value halfway between $U(f;P)$ and $L(f;P)$ as the best estimate. This number is $\frac{11}{32}$. The correct value is $\frac{1}{3}$, and the error is $\frac{11}{32} - \frac{1}{3} = \frac{1}{96}$. ■

It is intuitively clear that the upper sum *over estimates* the area under the curve, and the lower sum *under estimates* it. Therefore, the expression $\int_a^b f(x)\,dx$, which we are trying to define, is *required* to satisfy the basic inequality

$$L(f;P) \le \int_a^b f(x)\,dx \le U(f;P) \tag{A.1}$$

for all partitions P. It turns out that if f is a *continuous* function defined on $[a, b]$, then Inequality (1) does indeed define the integral. That is, there is one and only one real number that is greater than or equal to all lower sums of f and less than or equal to all upper sums of f. This unique number (depending on f, a, and b) is defined to be $\int_a^b f(x)\,dx$. The integral also exists if f is monotone increasing on $[a, b]$ or monotone decreasing on $[a, b]$.

Riemann-Integrable Functions

We consider the least upper bound (supremum) of the set of all numbers $L(f; P)$ obtained when P is allowed to range over all partitions of the interval $[a, b]$. This is abbreviated $\sup_P L(f; P)$. Similarly, we consider the greatest lower bound (infimum) of $U(f; P)$ when P ranges over all partitions of $[a, b]$. This is denoted by $\inf_P U(f; P)$. Now if these two numbers are the same—that is, if

$$\inf_P U(f; P) = \sup_P L(f; P)y \tag{A.2}$$

then we say that f is **Riemann-integrable** on $[a, b]$ and define $\int_a^b f(x)\,dx$ to be the common value obtained in Equation (2). The important result mentioned above can be stated formally as follows:

Riemann Integral Theorem. *Every continuous function defined on a closed and bounded interval of the real line is Riemann-integrable.*

There are plenty of functions that are *not* Riemann-integrable. The simplest is known as the **Dirichlet function**:

$$d(x) = \begin{cases} 0 & \text{if } x \text{ is rational} \\ 1 & \text{if } x \text{ is irrational} \end{cases}$$

For any interval $[a, b]$ and for any partition P of $[a, b]$, we have $L(d; P) = 0$ and $U(d; P) = b - a$. Hence, we obtain

$$0 = \sup_P L(d; P) < \inf_P U(d; P) = b - a$$

In calculus, it is proved not only that the Riemann integral of a continuous function on $[a, b]$ exists but also that it can be obtained by two limits:

$$\lim_{n\to\infty} L(f; P_n) = \int_a^b f(x)\,dx = \lim_{n\to\infty} U(f; P_n)$$

in which $P_0, P_1, \ldots$ is any sequence of partitions with the property that the length of the largest subinterval in P_n converges to zero as $n \to \infty$. Furthermore, if it is so arranged that P_{n+1} is obtained from P_n by adding new points (and not deleting points), then the lower sums converge *upward* to the integral and the upper sums converge *downward* to the integral. From the numerical standpoint, this is a desirable feature of the process because at each step, an interval that contains the unknown number $\int_a^b f(x)\,dx$ will be available. Moreover, these intervals shrink in width at each succeeding step.

Examples and Pseudocode

The process just described can easily be carried out on a computer. To illustrate, we select the function $f(x) = e^{-x^2}$ and the interval $[0, 1]$; that is, we consider

$$\int_0^1 e^{-x^2}\, dx \tag{A.3}$$

This function is of great importance in statistics, but its indefinite integral cannot be obtained by the elementary techniques of calculus. For partitions, we take equally spaced points in $[0, 1]$. Thus, if there are to be n subintervals in P_n, then we define $P_n = \{x_0, x_1, \ldots, x_n\}$, where $x_i = ih$ for $0 \le i \le n$ and $h = 1/n$. Since e^{-x^2} is *decreasing* on $[0, 1]$, the least value of f on the subinterval $[x_i, x_{i+1}]$ occurs at x_{i+1}. Similarly, the greatest value occurs at x_i. Hence, $m_i = f(x_{i+1})$ and $M_i = f(x_i)$. Putting this into the formulas for the upper and lower sums, we obtain for this function

$$L(f; P_n) = \sum_{i=0}^{n-1} hf(x_{i+1}) = h\sum_{i=0}^{n-1} e^{-x_{i+1}^2}$$

$$U(f; P_n) = \sum_{i=0}^{n-1} hf(x_i) = h\sum_{i=0}^{n-1} e^{-x_i^2}$$

Since e^{-x^2} is *decreasing* on $[0, 1]$, the least value of f on the subinterval $[x_i, x_{i+1}]$ occurs at x_{i+1}. Similarly, the greatest value occurs at x_i. Hence, $m_i = f(x_{i+1})$ and $M_i = f(x_i)$. Putting this into the formulas for the upper and lower sums, we obtain for this function

$$L(f; P_n) = \sum_{i=0}^{n-1} hf(x_{i+1}) = h\sum_{i=0}^{n-1} e^{-x_{i+1}^2}$$

$$U(f; P_n) = \sum_{i=0}^{n-1} hf(x_i) = h\sum_{i=0}^{n-1} e^{-x_i^2}$$

Since these sums are almost the same, it is more economical to compute $L(f; P_n)$ by the given formula and to obtain $U(f; P_n)$ by observing that

$$U(f; P_n) = hf(x_0) + L(f; P_n) - hf(x_n) = L(f; P_n) + h(1 - e^{-1})$$

The last equation also shows that the interval defined by Inequality (1) is of width $h(1 - e^{-1})$ for this problem.

Here is a pseudocode to carry out this experiment with $n = 1000$:

```
program Sums
integer i;   real h, sum, sum_lower, sum_upper, x
integer n ← 1000;   real a ← 0, b ← 1
h ← (b − a)/n
sum ← 0
for i = n to 1 step −1
    x ← a + ih
    sum ← sum + f(x)
end for
sum_lower ← (sum)h
sum_upper ← sum_lower + h[f(a) − f(b)]
output sum_lower, sum_upper
end program Sums

real function f(x)
real x
f ← e^(−x^2)
end function f
```

A few comments about this pseudocode may be helpful. First, a subscripted variable is not needed for the points x_i. Each point is labeled x. After it has been defined and used, it need not be saved. Next, observe that the program has been written so that only one line of code must be changed if another value of n is required. Finally, the numbers $e^{-x_i^2}$ are added in order of *ascending* magnitude to reduce roundoff error. However, roundoff errors in the computer are negligible compared to the error in our final estimation of the integral. This code can be used with any function that is *decreasing* on $[a, b]$ because with that assumption, $U(f;P)$ can be easily obtained from $L(f;P)$. (See Exercise A.3.)
The computer program corresponding to the pseudocode produces as output the following values of the lower and upper sums:

$$sum\_lower = 0.74651, \qquad sum\_upper = 0.74714$$

Example. If the integral

$$\int_0^{\pi} e^{\cos x}\,dx$$

is to be computed with absolute error less than $\frac{1}{2} \times 10^{-3}$, and if we are going to use upper and lower sums with a uniform partition, how many subintervals are needed?

Solution. The integrand, $f(x) = e^{\cos x}$, is a decreasing function on the interval $[0, \pi]$. Hence, in the formulas for $U(f;P)$ and $L(f;P)$, we have

$$m_i = f(x_{i+1}), \qquad M_i = f(x_i)$$

Let P denote the partition of $[0, \pi]$ by $n + 1$ equally spaced points,

$$0 = x_0 < \cdots < x_n = \pi$$

Then there will be n subintervals, all of width π/n. Hence, we have

$$L(f;P) = \frac{\pi}{n}\sum_{i=0}^{n-1} m_i = \frac{\pi}{n}\sum_{i=0}^{n-1} f(x_{i+1}) \tag{A.4}$$

$$U(f;P) = \frac{\pi}{n}\sum_{i=0}^{n-1} M_i = \frac{\pi}{n}\sum_{i=0}^{n-1} f(x_i) \tag{A.5}$$

The correct value of the integral lies in the interval between $L(f;P)$ and $U(f;P)$. We take the *midpoint* of the interval as the best estimate, thus obtaining an error of at most

$$\frac{1}{2}[U(f;P) - L(f;P)]$$

that is, the length of half the interval. To meet the error criterion imposed in the problem, we must have

$$\tfrac{1}{2}[U(f;P) - L(f;P)] < \tfrac{1}{2} \times 10^{-3}$$

From Formulas (4) and (5), we can calculate the difference between the upper and lower sums. This leads to $(\pi/n)(e^1 - e^{-1}) < 10^{-3}$. With the aid of a calculator, we determine that n must be at least 7385. ■

Summary

- Let $P = \{a = x_0 < x_1 < x_2 < \cdots < x_{n-1} < x_n = b\}$ be a **partition** of the interval $[a,b]$, which divides the interval $[a,b]$ into n subintervals $[x_i, x_{i+1}]$. The **lower sums** and **upper sums** of f corresponding to the given partition P are

$$L(f;P) = \sum_{i=0}^{n-1} m_i(x_{i+1} - x_i)$$
$$U(f;P) = \sum_{i=0}^{n-1} M_i(x_{i+1} - x_i)$$

  where m_i is the **greatest lower bound** and M_i is the **least upper bound** of $f(x)$ on the subinterval $[x_i, x_{i+1}]$, namely,

$$m_i = \inf\{f(x) : x_i \le x \le x_{i+1}\}$$
$$M_i = \sup\{f(x) : x_i \le x \le x_{i+1}\}$$

- We have

$$L(f;P) \le \int_a^b f(x)\,dx \le U(f;P)$$

Exercises

A.1. If we estimate $\int_0^1 dx/(x^2+2)$ by means of a lower sum using the partition $P = \{0, \frac{1}{2}, 1\}$, what is the results?

A.2. What is the results if we estimate $\int_1^2 dx/x$ by means of the upper sum using the partition $P = \{1, \frac{3}{2}, 2\}$?

A.3. For a decreasing function $f(x)$ over an interval $[a, b]$ with n uniform subintervals, show that the difference between the upper sum and lower sum is given by the expression

$$[(b-a)/n][f(a)-f(b)]$$

A.4. (Continuation) Repeat the exercise for an increasing function.

A.5. If upper and lower sums are used with regularly spaced points to compute $\int_2^5 dx/\log x$, how many points are needed if one is to achieve an accuracy of $\frac{1}{2}\times 10^{-4}$?

A.6. Let f be an increasing function. If the integral $\int_0^1 f(x)\,dx$ is to be estimated by using the method of upper and lower sums, taking n equally spaced points, what is the relationship between $\int_a^b f(x)\,dx$ and $\int_\alpha^\beta f^{-1}(x)\,dx$.

A.7. Find the relationship between the midpoint rule and the upper and lower sums. (See Exercise 5.1.6–7 in the textbook.)

Computer Exercises

A.1. Write a general-purpose procedure to estimate integrals of decreasing functions by the method of upper and lower sums with a uniform partition. Give the procedure with the calling sequence

real function `Integral`$(f, a, b, \varepsilon, n, \mathit{sum\_lower}, \mathit{sum\_upper})$

where f is the function name, a and b are the endpoints of the interval, and ε is the tolerance. The procedure determines n so that

$$\mathit{sum\_upper} - \mathit{sum\_lower} < 2\varepsilon$$

The procedure returns the average of *sum_upper* and *sum_lower*. Using $\varepsilon = \frac{1}{2}\times 0^{-5}$, test it on the **sine integral**

$$\mathrm{Si}(x) = \int_0^x \frac{\sin t}{t}\,dt$$

Hint: $\mathrm{Si}(1) \approx 0.94608\,27$ and

$$\mathrm{Si}(x) \approx x - \frac{x^3}{18} + \frac{x^5}{600} - \frac{x^7}{35280}$$

See Computer Exercise A.1.

Solutions Exercises

A.1. $L(f;P) = \sum_{i=0}^{n-1} m_i(x_{i+1} - x_i) = \frac{1}{\frac{1}{4}+2}\left(\frac{1}{2}\right) + \frac{1}{1+2}\left(\frac{1}{2}\right) = \frac{7}{18} \approx 0.38889$

A.5. Since $f(x) = 1/\log x$ is a decreasing function on the interval $[2,5]$, we can use Exercise A.6. If we estimate the integral by the average of the upper and lower sums, the error will be at most

$$\frac{3}{2n}\left[\frac{1}{\log 2} - \frac{1}{\log 5}\right] < \frac{1}{2}10^{-4}$$

Thus, we have $n \geq 56738$.

A.6. By Exercise A.4, we have $U - L = \frac{1}{n}[f(1) - f(0)]$ which is the worst possible error.

A.7. Clearly, we have

$$m_i = \min_{x_{i-1} \leq x \leq x_i} f(x) \leq f\left(\frac{1}{2}(x_i + x_{i-1})\right) \leq \max_{x_{i-1} \leq x \leq x_i} f(x) = M_i$$

Hence, we obtain

$$\begin{aligned} L(f;P) &= \sum_{i=1}^{n} m_i(x_i - x_{i-1}) \leq \sum_{i=1}^{n}(x_i - x_{i-1}) f(\frac{1}{2}(x_i + x_{i-1})) \\ &\leq \sum_{i=1}^{n} M_i(x_i - x_{i-1}) = U(f;P) \end{aligned}$$

Thus, we find

$$L(f;P) \leq M(f;P) \leq U(f;P)$$

Appendix B

Representation of Numbers in Different Bases

Exercises B.1

B.1.1. **a.** $e \approx (2.718)_{10} = (2.55747\ldots)_8 = (010.101\ 101\ 111\ 100\ 111\ldots)_2$
$\approx\ (2.5575)_8 = 2 + 2941/4096 \approx 2.718$

B.1.2. **d.** $(23.58)_{10} = (27.45075\ 34121\ 7\ldots)_8$

e. $(75.232)_{10} = (113.16662\ 13207\ 1\ldots)_8$

f. $(57.321)_{10} = (71.24426\ 41625\ 4\ldots)_8$

B.1.3. **a.** $(110\ 111\ 001.101\ 011\ 101)_2 = (\texttt{1B9.AE8})_{16} = (671.535)_8$????
$= 6 \times 8^2 + 7 \times 8 + 1 + 5 \times 8^{-1} + 3 \times 8^{-2} + 5 \times 8^{-3}$
$= 441 + 349/512 = 441.68164\ 0625$

b. $(001\ 001\ 100\ 101.011\ 010)_2 = (265.68)_{16} = (1145.32)_8$
$= 5 + 4 \times 8 + 1 \times 8^2 + 1 \times 8^3 + 3 \times 8^{-1} + 2 \times 8^{-2} = 613 + 26/64 = 613.40625$

B.1.4. **c.** $(47)_{10} = (57)_8 = (101\ 111)_2$

e. $(51)_{10} = (63)_8 = (110\ 011)_2$

g. $(011\ 011.111\ 010\ 110\ 110\ 100)_2 = (33.72664)_8 = (27.91956)_{10}$

B.1.6. $(0.4)_{10} = (0.3146\ 3146\ldots)_8 = (0.011\ 001\ 100\ 110\ldots)_2$

B.1.9. $(a_n a_{n-1} \ldots a_1 a_0)_{10} = a_n \cdot 10^n + a_{n-1} \cdot 10^{n-1} + \cdots + a_1 \cdot 10 + a_0$
$= a_0 + 10(a_1 + 10(a_2 + \cdots + 10(a_{n-2} + 10(a_{n-1} + 10(a_n)) \cdots))$

Each a_i is an integer in $\{0, 1, 2, \ldots, 9\}$ which could be expressed in binary as could 10. The calculations would be performed in binary arithmetic.

$$\begin{aligned}(479)_{10} &= 4 \times 10^2 + 7 \times 10 + 9 = 9 + 10(7 + 10(4)) \\ &= 1001 + 1010(111 + 1010(100)) = 1001 + 1010(111 + 101000) \\ &= 1001 + 1010(101111) = 1001 + 111010110 = (111011111)_2\end{aligned}$$

Check: $(111\ 011\ 111)_2 = (737)_8 = (479)_{10}$

B.1.12. ($\Rightarrow$) Suppose that a real number R has a finite representation in binary system $R = (a_m a_{m-1} \ldots a_1 a_0 . b_1 b_2 \ldots b_n)_2$.
Then $R = (a_m \ldots a_1 a_0 b_1 b_2 \ldots b_n)_2 \times 2^{-n} = m \times 2^{-n}$ where $m = (a_m \ldots a_1 a_0 b_1 b_2 \ldots b_n)_2$.

($\Leftarrow$) Suppose $R = m2^{-n}$.
Let $m = (a_m a_{m-1} \ldots a_1 a_0 b_1 \ldots b_n)_2$.
Then $R = (a_m a_{m-1} \ldots a_1 a_0 b_1 \ldots b_n) \times 2^{-n} = a_m a_{m-1} \ldots a_1 a_0 b_1 b_2 \ldots b_n$.
Therefore, R has a finite representation in binary system.

Appendix C

MATLAB: Brief Introduction and Tips

MATLAB is software package for carrying out numerical manipulations and displaying visual results among many other things. It can be run interactively, via scripts, and in the background. Here are some elementary introductory material and some simple tips on using MATLAB.

C.1. To start MATLAB, type `matlab` or double-click on the MATLAB icon. To leave MATLAB, use the command `quit` or `exit`. Type commands in the window with the prompt `>>`

C.2. There are three separate display windows for commands, graphics, and editing. For example, you can type commands in the edit window, load and execute them in the command window, and see visual displays in the graphics windows.

C.3. A convenient way to use MATLAB is via an external file and using multiple windows. For example, you can create a file called `test.m` using a text editor and type MATLAB definitions into it and save it without closing the editor. Then start MATLAB in another window and at the prompt type: `>>test` (not including the file extension `.m`), which loads and executes this file.

C.4. In your computer work-space, it is a good idea to set up a directory (or *folder*) devoted to MATLAB codes and results. Also, you can type `more on` to arrest the display of output to the screen and type `more off` to turn this feature off. It can also be included in your `startup.m` file as discussed in the next item.

C.5. In the directory mentioned above, it is helpful to have a file `startup.m` that will be read and executed whenever you start up MATLAB. For example, your might have `diary on` in your `startup.m` file. This command starts up a record of your session in a file called `diary`. At the end of the session, you can rename `diary` to be `diary32` or whatever, if you wish to save it. You can also edit it and send the diary-file to the printer. The command `diary off` turns this feature off.

C.6. You can edit command lines in the Command window. Use `clc` to clear the Command window or Edit: Clear Command window.

C.7. A great deal of useful information is available at `http:www.mathworks.com` The Help window is most useful for beginners. Select Help and drag-and-drop on MATLAB Help to open the Help browser. See also Help: MATLAB: Getting Started: Introduction. To start up demos and other windows try Launch Pad: MATLAB: Demos.

C.8. In MATLAB, there are different types of files such as m-files, mat-files, and ascii-files. An *m-file* is one having `.m` as its extension; that is, the name `example7.m` could be used. If you are writing a complicated MATLAB program, it is a good idea to create such a file using an editor so that you do not have to retype all the commands each time you test it. Once you are within MATLAB, you can read in and execute this program by simply typing the file name such as `example7` in which `.m` is excluded.

C.9. In using the menu-driven version of MATLAB, use the toolbar at the top of the window to pull-down various options. For example, click on `File` and drag-and-drop on `New` to load a m-file from a directory. After loading the file, execute it by selecting the icon `Debugging` and then drag-and-drop on `Run`. We denote this as `Debugging:Run`.

C.10. Normally statements/command are terminated with the enter key; however, they can be continued to the next line with three periods `...` followed by a return. Several statements can be on the same line by separating them with commas or semicolons. The semicolon is useful in suppressing unwanted displays. MATLAB is case-sensitive with regard to commands, functions, and parameters. Use parentheses in expressions if they would be ambiguous otherwise.

C.11. The symbol `%` indicates that the remainder of the line is a comment.

C.12. To enter a matrix, follow this example: `A=[1 2 3; 4 5 6; 7 8 9]`
To designate new entries, use spaces or commas. The semicolon symbol `;` ends a row of the matrix. To display a matrix or a variable just type its name. For example: `>>A` You can use built-in functions such as `triu(A)` for the upper triangular part of a matrix, `tril(A)` for the lower triangular part of a matrix, and `rank(A)` for the rank of a matrix.

C.13. For the transpose of a vector `v` or matrix `A`, type `v'` or `A'`, using the prime symbol, which is next to the return/enter key on most keyboards.

C.14. To enter column vectors, follow these examples: `x=[1; 2; 3]` or
`y=[2/3 1+3 7^2]'` These are matrices with three rows and one column. Elements of matrices and vectors can be either numbers or expressions. To enter a row vector, follow these examples: `[1 2 3]` or `x'` or `y'`.

C.15. When dealing with vectors and matrices, the operations add `+`, subtract `-`, multiply `*` , divide `/`, and power `^` are computed element-wise/entry-wise by preceding them with a period such as `.+, .-, .*, ./, .^` These are particularly useful in graphics and throughout MATLAB because many variables are vectors and matrices. For example, compare

```
x=[1 1/2 1/3]*[1; 2; 3]
y=[1 1/2 1/3].*[1; 2; 3]
x*y
x.*x
y.*y
```

C.16. Special matrices are easily generated by built-in functions such as `ones(3)` for a 3×3 matrix of all 1's, `zeros(4,2)` for a 4×2 matrix of all 0's, `eye(5,5)` for a 5×5 identity matrix, `rand(2,4)` for a 2×2 matrix with uniformly distributed random entries, `hilb(3)` for a 3×3 Hilbert matrix, and many more.

Another example is to create a diagonal matrix by using `D=diag([1,3,7])`

C.17. The statement `x=[0:0.2:1]` creates an array with elements from 0 to 1 with increment of 0.2; namely, `x=[0 0.2 0.4 0.6 0.8 1]`. The limits and the step size can be numbers or expressions.

C.18. By itself, the colon denotes an entire row or column. For example, if `A = rand(5,5)` then `A(1:3, 2)` is the column vector consisting of the first to third entries of the second column of array `A`, `A(1:5, 5)` is the fifth column, `A(1:3, 1:5)` is the first three rows, `A(:, [1 3])` contains the first and third columns.

C.19. While all computations in MATLAB are carried out in double precision, the displayed output varies according to these commands, which stay in effect until changed: `format short` for 4 decimal places in fixed point (default), `format short e` for 4 decimal places in scientific notation, `format long` for 14 decimal places in fixed point, `format long e` for 15 decimals places in scientific notation, `format rat` for approximation with a ratio of small integers. Hence, to see more digits in the computed results type: `format long`

C.20. A useful function to print output is `disp`. Examples are
`disp('The vector x is')` or `disp([1 2 3])` or
`disp(['Convergence after', num2str(iter), 'iterations'])`
where `num2str` converts a number to a string. Another more flexible function is `fprintf` that has the form `fprintf('filename','format',list)` If the filename is omitted the output is to the screen. Common formats are `%M.Nf` for fixed point, `%M.Ne` for floating point, `%M.Ng` for automatic selection of fixed or floating point. Here `M` is the field width and `N` is the number of decimal places. The list contains variable names separated by commas together with `\n` that produces a new line. For example,
`k = 5; x = pi; fprintf('\n k=%1.0f\n pi =%15.14f',k,x)`

C.21. Some miscellaneous commands are: `clc` clears the command window, `clf` clears the current figure window, `what` lists all the MATLAB-related files in the directory, `which` locates functions and files, `who` list current varialbes, `whos` gives detailed information about existing variables, `why` answers questions.

C.22. To compute the matrix-vector multiplication Ax, type: `A*x`

C.23. There is a vast collection of built-in functions from elementary ones such as `abs(x)`, `sin(x)`, `exp(x)`, `log(x)`, to more sophisticated functions such as `inv(A)` for matrix inversion,, `eig(A)` for eigenvalues/eigenvectors, `svd(A)` for singular-value decomposition, and many more. All function names are lower case letters and the functions have single or multiple arguments.

C.24. To call for the row-reduced echelon form of a matrix, follow this example: `rref(A)` or, if you want to save the results and use it later then type: `B=rref(A)`

C.25. To see the row reduction algorithm applied to a matrix A, step-by-step, follow this example: `rrefmovie(A)`

C.26. To get the determinant of A follow this example: `det(A)` or `t=det(A)` if you will need the value later.

C.27. To get the characteristic polynomial of a matrix A follow this example: `poly(A)` or `p=poly(A)`. This delivers the coefficients of the characteristic polynomial, starting with the coefficient of the highest power of the variable.

C.28. To get the matrix P of eigenvectors and the diagonal matrix D containing the eigenvalues of A the command `[P,D]=eig(A)` could be used. The eigenvectors of A are the columns of P ordered just as the eigenvalues are ordered in D. MATLAB always gives eigenvectors as unit vectors. Thus, an eigenvector x is delivered as $x/||x||$. Here we have found the factorization: $A = PDP^{-1}$.

C.29. To compute the Euclidean norm of a vector x use `norm(x)` or `t=norm(x)`. Other norms are invoked with `norm(x,1)`, `norm(x,2)`, and `norm(x,inf)`.

C.30. To get the inverse of A use: `inv(A)` or `B=inv(A)`

C.31. To form a matrix B consisting of rows 1 to 6 and columns 1 to 7 taken out of A, follow this example: `B=A(1:6,1:7)`

C.32. To form a matrix B consisting of columns 4 to 6 out of the 3×6 matrix A, follow this example: `B=A(1:3,4:6)`

C.33. To get the LU-factorization of a matrix A, use: `[L,U]=lu(A)`
in which L is a lower triangular matrix with 1s on its diagonal, and U is an upper triangular matrix. The relationship is $A = LU$.

C.34. To get the LU-factorization of a permuted matrix use: `[L, U, P]=lu(A)`
The resulting matrices L, U, P have this relation: $PA = LU$, where P is a permutation matrix. Thus, PA has the rows of A in permuted order.

C.35. To solve the equation Ax=b, one has various methods available in MATLAB. First, one can use `rref(B)` on the augmented matrix $B = [A \mid b]$. Second, one can compute `C=inv(A)` and use `x=C*b`. This is *not* the recommended method! Thirdly, one can use the special MATLAB command `x=A\b`. This solves the equation in an efficient manner, probably computing first the LU-factorization of A and then solving $Ly = b$ and $Ux = y$.

C.36. To get the condition number of A, use: `cond(A)`

C.37. To produce the Cholesky factorization of A, use: `R=chol(A)`
Then $A = R^T R$, where R is an upper triangular matrix. Here the matrix A should be symmetric and positive definite.

C.38. To get a QR-factorization of a matrix A, use `[Q,R]=qr(A)`. Here $A = QR$ where Q is a unitary matrix and R is an upper triangular matrix. The dimensions are compatible such as $A_{m\times n}$, $Q_{m\times m}$, $R_{m\times n}$.

C.39. To get the singular-value decomposition of A use: `[U,S,V]=svd(A)`
Here $A = USV$ where U and V are unitary matrices, and S is a diagonal matrix having the singular values of A on its diagonal. Here the dimensions are $A_{m\times n}$, $U_{m\times m}$, $S_{m\times n}$, $V_{n\times n}$.

C.40. MATLAB has its own programming language with a variety of flow control statements such as `for`, `if`, `while`, and so on. Relational operators are `<=` for $\leq$, `<` for $<$, `>=` for $\geq$, `>` for $>$, `==` for $=$, `~=` for $\neq$. Notice that `==` is a relation while `=` is used in an assignment statement. Logical operators are `&` for and, `|` for or, `~` for not. A simple example of an if-statement is:

```
if i<=j
  A(i,j)=1;
else
  A(i,j)=0;
end
```

Note that vector and matrix indices begin at 1. An example of nested for-loops is

```
for i=1:5
  for j=1:5
    H(i,j) = 1/(i+j+1);
  end
end
```

This can be computed more efficiently using `hilb(5)`. Here is an example of a while-loop

```
n = 0
while 2^n <= 1e5
  n = n+1;
end
n
```

This computes the smallest nonnegative integer n such that $2^n > 10^5$,which can be computed more efficiently using `[f,n] = log2(1e5)`.

C.41. A users define function for $f(x) = e^x - \cos(2x)$ can written in an m-file called `fcn2.m` of the form

```
function y=fcn2(x)
y = exp(x) - cos(2.*x);
```

To write and edit m-files, you can use any editor or the Editor/Debugger in MATLAB. Then within the command window the function can be evaluated such as `>>fcn2(3.5)` Functions can have other functions as argument such as `r = fzero(@fcn2, 1)` which finds the zero of the indicated function closest to 1. To plot the graph of the function, use `fplot(@fcn2, [-1,1]);` or
`fplot('exp(x)-cos(2.*x)', [-1,1]);`

C.42. Some file I/O functions are: `fclose` closes a file, `fopen` opens a file, `fgets` reads a line from a file, `fprintf` writes formatted data to a file, `fscanf` reads formatted data from a file, `feof` writes formatted data to a file, `fclose` closes a file, `fclose` closes a file,

C.43. Plots of 2D functions can be generated with the command `plot`. For example, `x = [-1:0.1:1]; y = x.^2; plot(x,y)` gives the graph of the function $y = x^2$ over the interval $[-1, 1]$. If the function is defined in an m-file, say `fcn1.m` then the command `fplot(@fcn1, [-1,1])` plots the function over the indicated interval. Commands such as `grid`, `xlabel`, `ylabel`, `title`, and so on can be used to add or adjust various features of a plot.

C.44. An example of multiple plots on a single graph is `x = [-1:0.1:1];` `f1 = exp(x); f2 = sin(2.*x); plot(x,f1,x,f2)` Similar commands can be used for plotting parametrically defined curves.

C.45. Mesh 3D perspective surface plots can be drawn with the function `mesh`. For example, `[x,y] = meshgrid(-5:0.2:5, -5:0.2:5);` `z = 1./exp(x.^2 + y.^2); mesh(x,y,z)` generates the plot of the function $z = 1/e^{x^2+y^2}$ over the square $[-5, 5] \times [-5, 5]$. Other useful 3D graphics commands are `plot3` and `surf`.

C.46. We recommend using (a) periods before operations `+,-,*, /, ^` to make sure they are done elementwise, (b) periods in numbers to make sure they are done in floating-point, and (c) parenthesis around things that may be ambiguous; for example,
`f=(1.0+(3.0/4.0).*x.^2)./(2.0.*x.^3-(5.0/9.0).*x.^5)`

C.47. The command `subplot(m,n,i)` creates `mn` plots in `m` rows and `n` columns where the next plot goes in the `i`-th position (counting left-to-right across the columns and down the rows). As an illustrations of some of the various options available, the following plot commands produce two side-by-side plots with the first one containing two curves: sine as a dashed line, cosine as a dotted line, and the nodal points are marked as green o's and magenta x's.

```
x = [0: 0.1 : 2*pi];
subplot(1,2,1)
plot(x,sin(x),'--',x,sin(x),'go',x,cos(x),':',x,cos(x),'mx')
subplot(1,2,2)
plot(x, 1./(1+x.^2))
```

Note the use of periods in the last plot to ensure that the correct (element-wise) operations are carried out for the vector x.

C.48. Two other ways of producing multiple plots on a single graph
`x = [0: 0.1 : 2*pi]; y1=sin(x); y2=sin(2.*x); plot(x,y1,y2)` or
`x = [0: 0.1 : 2*pi]; Y=[sin(x)', sin(2.*x)']; plot(x,Y)`

C.49. Yet another example is the cosine curve and two partial sums from its Maclaurin series
`x = [0: 0.1 : 2*pi]; plot(x, cos(x), x, 1-x.^2./2, x, 1-x.^2./2+x.^4./24)`
Note the use of the operations elementwise power `.^` and elementwise divide `./`

C.50. The command `print -dps graph.ps` produces a file `graph.ps` in your computer. On a Unix computer, this file can be printed on laser printer number 31 by typing `lpr -Plw31 graph.ps` (You may also use `.eps` for encapsulated postscript.)

C.51. The plot-window can also be used directly to generate and print plots.

C.52. The plot command `axis([-6 6 -2 2])` overrides auto-scaling and produces a plot window of size $-6 < x < 6$ and $-2 < y < 2$. By default, the axes are auto-scaled. The command `axis` freezes the current scaling for subsequent plots and typing `axis` again returns you to auto-scaling. The command `axis('square')` results in the same scale being used on both axes.

C.53. The command `hold` freezes the graphics screen so that subsequent plots are superimposed on the same plot until release by typing the command `hold` again. The commands `hold on` and `hold off` may also be used.

C.54. Some of the plot commands for line types, mark types, and colors: linetypes: dashes (`--`), dotted (`:`), dashdot (`-.`), solid (`-`) [default]; marktypes: point (`.`), plus (`+`), star (`*`), circle (`o`), x-mark (`x`), square (`s`), diamond (`d`), up-triangle (`v`), down-triangle (`^`), left-triangle (`<`), right-triangle (`>`), pentagram (`p`), hexagram (`h`); colors: yellow (`y`), magenta (`m`), cyan (`c`), red (`r`), green (`g`), white (`w`), black (`k`), blue (`b`) [default].

C.55. Some of the commands for labeling plots:
`title("This is the title of the plot.")`, `xlabel("x-axis")`, `ylabel("y-axis")`, `text` – position text at specific coordinates, `gtext` – interactively-positioned text (allows mouse or arrow key to position text at a cross-hair on the graph).

Appendix D

Maple: Brief Introduction and Tips

Maple is a software package for both symbolic and numeric computations.

D.1. Double click the MAPLE icon to obtain a window version with menus, which opens up a new window for input and output. To open additional new windows, click on `File` and drag-and-drop on `NEW` or click on the icon with a blank page on it in the upper lefthand corner. An Unix X-windows version containing graphics is `xmaple`. Another Unix version of Maple can be invoked by typing `maple` at the Unix command prompt. It can be used from a dumb terminal connection for many applications not requiring a graphical display.

D.2. The Maple prompt is `>` in both the Windows and Unix versions. You can type `?` to bring up the first help screen for Maple. If you are already familiar with Maple syntax, you can type `?intro` to get some initial information on how to get started with Maple. In `xmaple`, click and open the `?Help` window to learn about Maple. Anytime you need help on a particular topic, you can look it up in `Help` by typing `?topic`. To get out of `Help`, press `q` to return to the Maple command prompt. Alternatively in Windows, you can go to `Help` on the menu bar to get more information. For a great deal of information on Maple with many example files as well as an online tutorial, visit the Maple homepage at `http://www.maplesoft.com/flash/index.html`

D.3. In the Windows version of Maple, many of the buttons are quite useful. For instance, on the lefthand side of the menu toolbar is an icon of a large check that can be used to check and to correct the syntax of an expression. The button next to it with an exclamation symbol on it executes the current expression. Then the button that looks like three exclamation points executes the entire Maple worksheet.

D.4. To obtain a numerical value for an expression, the function `evalf` is often useful. Floating-point evaluations can be done such as `evalf(2/3);` or `evalf(Pi,15);` to obtain π to 15 decimal digits. The default precision is 10 but it can be changed by giving `evalf` a second argument as in the example `evalf(exp(1),25);` which specifies that the displayed precision is 25. Alternately, the statement `Digits := 50:` changes the precision to fifty decimal digits for all of the following statements.

D.5. In a line, everything following `#` is a comment. The symbol `%` is the ditto operator, which refers back to the previous line, `%%` refers back two lines, and `%%%` refers back three lines, which is the limit for this symbol. Another useful feature of Maple is `evalf(%)` to get a numerical value of the previous expression. Likewise, `evalf(%%)` can be used to obtain the numerical value of the second previous express and `evalf(%%%)` for the third previous expression.

D.6. Output is displayed for commands ending with semicolon (`;`) but it is suppressed for commands ending with colon (`:`). If you forget to type either of these before going to the next line, then just type one of them at the beginning of the new line.

D.7. You can edit previous lines using high-lighting, cut-and-pasting, and many other editing features. Click on the `Edit` icon or use `Esc`-key to invoke the `vi` editor. If you are using a nongraphics version of Maple, type `?editing` for details.

D.8. Help can be obtained by clicking `Help` in the menu window or typing: `?`, `?intro`, `?help`, `?help[keyword]`, `?topic`, `index[category]`, and so on. Useful help screens are `series`, `integration,numerical`, `nextprime`, `solve`, and others. The commands `factor` and `simplify` are often useful.

D.9. The operations for addition $+$, subtraction $-$, multiplication $*$, division $/$, and exponentiation `**` or `^` are all carried out in *exact* arithmetic unless otherwise specified.

D.10. Maple contains all of the standard mathematical functions such as `abs`, `sqrt`, `exp`, `sin`, `cos`, and so on. Special mathematical constants are `Pi` for π, `I` for the complex number i, and `exp(1)` for e, and many others. When specifying the mathematical constant e, it is no longer possible to use `E` so you must use `exp(1)`.

D.11. The exclamation mar (`!`) means factorial so that `5!` means $5 \cdot 4 \cdot 3 \cdot 2 \cdot 1$.

D.12. Double dots are used for a range such as in `1..10` to signify the range $1, 2, 3, 4, 5, 6, 7, 8, 9, 10$. We can verify that $\sum_{i=1}^{n} i^2 = \frac{1}{6}n(n+1)(2n+1)$ by using `sum('i^2', 'i'=1..n);` followed by `factor(%);` Moreover, double dots are used to specify the range in plotting, such as in `plot(sin(x^3), x=-2..2);`.

D.13. Use := for an assignment statement in which a variable is given a value such as `x := 2.5;` or defining a system of equations such as `eqns := {2x-4y=5, 4x+7y=9};` which can be solved for the two variables by using `solve(eqns,{x,y});` An assignment can be undone with `unassign(varname)`. You can also do a `restart:` to remove all definitions at once. For example, if a letter has been assigned a value by a statement such as `j := 7;` and if later you want `j` to be a variable, then use `'x'`. Here is concrete example `j := 6; sum('2*j', 'j'=1..5);` This left-leaning single prime ' is on the key to the left of the key with the number 1 on it.

D.14. In Maple, there are some keywords that are reserved and can not be used or assigned as variable names since they are internal to Maple. Some examples of Maple keywords are and, or, stop, try, save. Also, Maple protects some letters such as D since it is used in D(x^2-2x+5) to find the derivative of the indicated function. Also, O is protected because it is reserved for the the Order function call. To check if a certain letter or word or expression is protected, type type(<name>, protected), which returns true or false.

D.15. A procedure or mapping is defined such as g := x->x**5 + 2*x - 5 to mean $g(x) = x^5 + 2x - 5$. To define a function, several other syntaxes can be used, illustrated here: f := 7.2*exp(x); f := 3*cos(x); f := x->7.2*tan(x);
f := (x,y)->cos(x+y);
f := proc(x) if x>2 then x^3 else end;

D.16. The symbol $ is the repetition operator; for example, diff(g(x), x$4); means the fourth derivative for the function $g(x)$.

D.17. Use =, <, <=, <>, >=, > for equal, less than, less than or equal, and so on; namely, $=, <, \leq, \neq, \geq, >$.

D.18. In Maple, programs can be written that contain *do-loops*.

```
sum = 0;
for i from 3 by 2 while i<15 do
  sum := sum + i
end do;
```

Note that there is only one instance of the character ; at the end of the loop, which is signified by end do.

D.19. Maple has many software packages that are quite useful. For carrying out linear algebra operations, the packages linalg and LinearAlgebra are similar. While LinearAlgebra is newer, linalg has many more routines. When dealing with matrices, you may need a function that is part of one if these packages. To link to the latter package, type with(LinearAlgebra): and then proceed to use functions that are contained within it such as Eigenvectors, Eigenvalues, GenerateMatrix, GenerateMatrix, MatrixInvers, GaussianElimination, and so on.

D.20. To input a file of commands or a program, type read <file-name>; Typically, Maple worksheets end with the extension .mws

D.21. In Unix, to save your session type save <file-name> or click on Save Worksheet. In Windows, click on File and drag-and-drop on Save. By default, it will be saved in a file with the extension .mws. This extension means that it is written in the Maple *internal* form that only maple can read!! To save a file that you can read and edit, use some other endings such as .txt. To save values of individual variables, one could use save x,i,"temp.mws"

D.22. To end a session type quit but remember to save results before quitting if you want to use them again. In Windows, you can also go to File and drag-and-drop to Exit.

D.23. Between unrelated tasks, it is recommended to clear memory by typing `restart:` Also, it is a good idea to turn on `AutoSave`, which is under `File` in the Windows version of Maple, which is defaulted to be on.

The following are some examples of typical Maple commands.

D.24. To define and evaluate a function:

```
f := (x,y,z)->2*x^3-3*y+5*z^3:
f(1,2,3);
```

D.25. To plot in 2D or 3D in the Windows version of Maple, click on `Insert` and drag-and-drop on `Plot` and then choose `2D` or `3D`. A 2D plotting example is `plot(sin(x), x=-2+Pi..2*Pi);`

D.26. A column vector or a column of a matrix is defined with $<\ \ >$ while a row vector or a row of a matrix uses $[\ \ ]$

```
u := <1,2>; v := [3,4];
Vector[row](u);
Vector[column](v);
array([1,2]);
```

D.27. Define an array and a matrix and carry out linear algebra computations:

```
array([[4,1],[3,2]]);
with(LinearAlgebra):
A := Matrix([[4,1],[3,2]]);
detA := Determinant(A);
evalA := Eigenvalues(A);
evecA := Eigenvectors(A);
(P,L,U) := LUDecomposition(A);
```

D.28. Define a random matrix and extract a row vector and a column vector:

```
B := RandomMatrix(2,2);
col2B := Column(B,2);
row2B := Row(B,2);
whattype(col2B); whattype(row2B);
```

D.29. Using floating-point numbers, evaluate a matrix:

```
C := evalf(B);
```

D.30. Compute a dot-product of a row vector and column vector:

```
p1 := DotProduct(row2B,col2B);
p2 := Transpose(col2B)&*row2B;
```

D.31. Define assignment statements to form matrices and multiply them elementwise:

```
rowA:= i->Row(A,i):
colB:= j->Column(B,j):
AB1:Matrix([[rowA(1).colB(1), rowA(1).colB(2)],
            [rowA(2).colB(1), rowA(2).colB(2)] ]);
AB2 := A.B;
```

D.32. Solve a systems of linear equations for two different methods:

```
solve({x-2*y=6,2*x-3*y=8}, {x,y});
sys := [x-2*y=6,2*x-3*y=8]:
var := [x,y]:
GenerateMatrix(sys,var,augmented=true);
LinearSolve(%);
```

D.33. Computing $x - \sin x$ from the first ten terms of its Taylor series and comparing this to the value from built-in functions: specific point.

```
y := sum( '(-1)^{k-1}*x^{2*k+1}/(2*k+1)!', 'k'=1..10);
Digits := 50:
x := 1.9;
z := evalf(y);
t := x - sin(x);
d := z - t;
```

D.34. Using the Taylor series method, to solve an ODE at two points

```
ode := D(x)(t) = 1 + x(t)^2 + t^3;
f := dsolve({ode, x(1)=-4}, x(t), numeric);
f(1); f(2);
```

D.35. Plot $\sin x$ and its the partial sum of its Maclaurin series on the interval $[-\pi, \pi]$

```
f := x -> sin(x);
k := x - x^3/6 + x^5/120;
interface(plotdevice=postscript);
interface(plotoutput=sinplot.ps);
plot {{f,k}, -Pi..Pi, numpoints=601, title='Sin x and 3rd Partial Sum');
```

Here `sinplot.ps` is the name assigned in the program to the postscript file, which can be printed on a laser printer with the command `lpr` or viewed using Ghostview on a Unix System. Using `xmaple`, this can be done by clicking on `File` and then drag-and-drop on `Print`.

D.36. Plot the Runge function and compare to polynomial interpolation

```
interface(plotdevice=postscript):
interface(plotoutput=rungeplot.ps):
n := 13;
h := 10./(n-1);
x := array(1..n):
y := array(1..n):
for i from 1 to n
  do
    x[i] = -5*(i-1)*h;
    y[i] := 1./(1. + x(i)^2);
  end do;
interp(x,y,t):
plot(%, t=-5..t, title='Runge Example n=13'):
```

D.37. Solving an ODE and plotting

```
with(plots):
interface(plotdevice=postscript):
interface(plotoutput=odeplot.ps):
odeq :=  (D@@2)(x)(t) + D(x)(t) + x(t)^2 - 2*t =0;
init := x(0)=0, D(x)(0)=0.1;
soln := dsolve({odeq, init}, x(t), 'numeric');
plot1 := odeplot(soln, {t,x(t)}, 0..3, title='ODE Problem'):
soln(0);
soln(0.5);
soln(3.0);
plot(plot1);
```

Appendix E

Mathematica: Brief Introduction and Tips

E.1. If you are using the Unix version of Mathematica, type the command `math` to initiate a Mathematica session with the non-menu-driven system. On the other hand, type `mathematica` for the menu-driven system. (This may vary depending on the computer system you are using.) If you type `mathematica` and you get a message that the fonts are not properly installed just repeatedly type `Continue` until it opens up—all of the fonts may not be correct but it most likely will still work!

E.2. If you are not using the menu-driven version and want a record of the session on a Unix system, type the command `script` before entering Mathematica to initiate a file called `typescript` in which all details will be saved. After exiting Mathematica, you should type `exit` to terminate the saving of information into the file `typescript`, which can then be edited.

E.3. If you are using the menu-driven version of Mathematica, you can easily save your notebook by opening `File` and dragging-and-dropping on `Save`, which creates a notebook file with the extension `.nb` , or selecting `Save as Special` for many other types of files that Mathematica can handle such as those ending with `.txt` for plain text, `.mat` for matrix format, `.jpg` for an image file, `.ps` for a postscript file, and so on. A plain text file can be read by the user and edited with any editor. The default file extension is `.nb` and a notebook file can only be read by Mathematica and edited therein.

E.4. If you are using the Windows version of Mathematica, then, after opening this application, you should see a menu bar across the top, an empty window, and a palette. The top is the usual menu bar similar to those in other Windows applications. The window titled `Untitled-1` is where you type in the input and where you see the output. To the right, there is a palette of many common functions and symbols. The following sub-items are some additional details on the Windows version of Mathematica:

a. To save your work in Mathematica, go to `File` and then drag-and-drop on `Save`. Mathematica will automatically save into the file `*.nb` in the notebook format. To save in other formats, go to `File`, click on `Save as Special`, and then choose the format you want to use.

b. To open an existing file in Mathematica, go to `File`, click on `Open`, and then select the file you want to open within a particular directory.

c. In the menu-driven version of Mathematica, material can be typed in or specified via a palette for Greek letter and mathematical symbols. To use the palette, click on the symbol or mathematical function you wish to use. To change the palette or to find symbols not on the existing palette, go to `File` and then select `Palette`.

d. Mathematica defaults all matrices to standard matrix form. To change this, click on `Cell`, then go to `Default Input Format Type`, and choose the type of input format you wish to use.

e. To input a 2×2 matrix, click on the button in the palette if it is on the existing palette. However, if it not on the existing palette or if you wish to input a matrix other than a 2×2 matrix, go to `Input`, click on `Create Table/Matrix/Palette`, and a dialog box will appear. By typing in the appropriate information, you can create any size/type of matrix. Alternatively, you can create a table or a palette.

f. After typing in an expression to be evaluated, press `<shift>+<enter>` to evaluate the cell. Alternatively, you can evaluate the entire notebook. To do this, go to `Kernel`, then select `Evaluation`, and finally choose `Evaluate Notebook`.

g. For help with Mathematica or to get information on how to evaluate a certain mathematical function within Mathematica, go to `Help`. Within the `Help` browser, type in the topic you wish to look up.

E.5. In the non-menu-driven Mathematica, to input from a file called "`file7.m`" (previously created with an editor), type `math < file7` This will also work for the menu-driven version `mathematica` but it is better to click on `File` and drag-and-drop to `Open`. If you see your file listed, then just click on it or type its name into the small window. If either of these does not, then click on `Update` or `Advanced` and then on `Standard` and either click on the file name or type it into the box, using its correct subdirectory address.

E.6. To find documentation on Mathematica, search the Web. The main Mathematica site is `www.wolfram.com`, which contains voluminous information. For example, click on `Resource Library` and then either `Documentation` or `How-Tos` or `Tutorials` or `Examples`, which are all excellent sources of information. Within `Documentation`, you can click on `Built-in Functions`. At any time, you can do a search for specific information.

E.7. Once Mathematica has opened, you can get information about a specific topic by following these models: `?Eigensystem` `?Factor` `?MatrixExp` `?Plot`
To get the names of all Mathematica routines beginning with the three letters "`Int`" give the command: `?Int*`. In the menu-driven version on Mathematica, click on `Help` to obtain information from a pull-down menu.

E.8. It is a good idea to add the following lines to the file "`init.m`" in your Mathematica directory `.Mathematica`

```
$Post:=If[MatrixQ[#],MatrixForm[#],#]&
<<LinearAlgebra`Master`
```

The first line sets all matrices in standard matrix form and the second loads `LinearAlgebra`Master``, which must be typed without spaces. The left-leaning single quote is used, which is on the key next to the number 1.

E.9. In Mathematica, all built-in commands must begin with a capital letter. For example, to terminate a Mathematica session, type `Quit`, with capital "`Q`" since the command `quit` will not work. (If you have been keeping a record of the session in the file `typescript`, remember to use `exit` outside Mathematica to end that process. This remark pertains to Unix-based operations.) In the menu-driven Mathematica, click on `File` and then drag-and-drop on `Quit`.

E.10. Mathematica commands end with a space or line return. You can suppress the output by ending a command with a semi-colon (;).

E.11. The symbol `%` is shorthand for "the preceding output," and two of them would refer back two commands and so on. To refer to line number four, use `%4`. Thus, having computed `x=Cos[0.45]`, one can ask for the result to be squared by writing `%` ∧ 2. For another example, we can enter (after the Mathematica prompt) the expression `Exp[x] + Cos[x]` and then, at the next prompt enter `w = %`. The result will be `w = Exp[x]+Cos[x]`.

E.12. Continuing the previous example, enter `Solve[w == 0]`. This asks Mathematica to find the *roots* of the function `w` from above. The symbol == is a logical connective. Think of the above Mathematica instruction as the command "determine the values of `x` for which the assertion "`w=0`" is *true*.

E.13. For some basic arithmetic, try these (type them on separate lines by using the enter key):

```
7+5   3^10   Pi   N[Pi]   N[E]   N[E, 25]
```

Here `N` evaluates a number or an expression to the default machine precision, usually six decimal digits. The last command asks for twenty-five decimals in the numerical value of the mathematical constant e.

E.14. To insert *comments* in a file use this form: `(* This is my comment *)`. These can be used anywhere in a file, such as just after a command that needs some documentation or explanation.

E.15. In Mathematica, functions are paramount, and they require the bracket notation $[\cdots]$. All built-in Mathematica functions begin with a capital letter. Examples for you to try:

```
Sin[Pi/6]   (* evaluation *)
ExpIntegral[3.2] (* integral *)
```

E.16. To plot the curve $y = x^2$, type: `Plot[x∧2, {x,0,2}]`. Mathematica responds with a plot displayed on the terminal. The range of x is the interval $0 \le x \le 2$. If you are not using the menu-driven version of Mathematica, it may be a coarse plot. However, if you are using the menu-driven Mathematica, you will see a nice plot drawn within your notebook. For a high-resolution plot file that can be printed on paper, or viewed using Ghostview, see the next item.

E.17. For producing a plot file named "p", type:

```
F[x_] := (1/Exp[x^2])
curve = Plot[F[x], {x,-1,1}, DisplayFunction->Identity]
Display["p",curve,"EPS"]
```

The file encapsulated postscript file `p.eps` is created by the last line of the code. It can be sent to a printer or viewed using Ghostview.

E.18. Here is an example of a "while-loop". It computes and displays the real and imaginary parts of the function $(x + iy)^n$. (We let n run from 1 to 12):

```
n=1;  While[(n=n+1)<13, Print[ComplexExpand[(x+Iy)^n]]]
```

E.19. To get the indefinite integral $\int (1 + \sin x)^{-1}\, dx$ use this command:

```
Integrate[(1+Sin[x])^(-1), x]
FullSimplify[%]
```

Mathematica responds with `1 - Cos[x]/(1 + Sin[x])`.

E.20. To define your own function of two variables, follow this example:

```
f[x_,y_] := (x*y)^2*(x+y)
```

Notice that it is safer to use $*$ to signify multiplication. (See next item.)

E.21. A point of syntax: you can write `5x`, but not `xy`. If the product of `x` and `y` is intended, write `x y`, with the space, or `x*y`, which is much safer. If the space is omitted, Mathematica interprets `xy` as a single variable named "xy".

E.22. Here is a definite integral $\int_0^3 x(x - \frac{1}{2})(x - 1)\, dx$ to serve as a model:

```
f[x_] := x*(x-1/2)*(x-1)
Integrate[f[x], {x, 0, 3}]
```

The answer `9` is the response.

E.23. An example of entering a 3×3 matrix and asking for its eigenvalues:

```
A={{3, 2, 7}, {-3, 1, 4}, {6, 5, -4}}
Eigenvalues[%]
```

E.24. An example showing a polynomial and asking for its zeros (roots):

```
p=2x^3 + 7x^2 - 15x + 2
Solve[p[x]==0]
```

E.25. An example of a pair of differential equations to be solved:

```
sys={x'[t]-3y[t]==0, 3y'[t]+x[t]==0, x[0]== 3, y[0]==0}
DSolve[sys, {x[t], y[t]}, t]
```

Sometimes (although not in this example) Mathematica will provide a solution that can be simplified by the command `Simplify[%]`.

E.26. Mathematica can differentiate and integrate, as in these examples:

```
D[Sin[x]/ArcCosh[x], x]
Integrate[Tan[x]^y, x]
Integrate[Exp[3*x] Sin[x]^2, x]
```

E.27. Suppose you want to execute a Mathematica file, say `file7`, that you have already typed, such as the following sample file containing linear algebra commands.

```
<<LinearAlgebra`MatrixManipulation`
(*Note: line above is typed without spaces and with back-quotes.*)
MatrixForm[MatA={{1,3,7,-5}, {-2,3,-3,4}, {1,-3,-2,1}}]
(*Standard way of entering a matrix, row-by-row.*)
RowReduce[MatA];
MatrixForm[%]
MatrixForm[MatC=TakeColumns[MatA,3]]
x={{1},{4/5},{-6/5}}
MatC.x
MatInv=Inverse[MatC]
% //MatrixForm
MatC.MatInv
```

Click on `File` and drag-and-drop on `Open`. When the small window opens, click on the file name if it is present or click on `Update` or `Advanced` otherwise. In the dialog box, type the file name including its directory, if different from the current one. To execute the file, click on `Kernel` and then on `Evaluation` followed by either `Evaluate Cell` or `Evaluate Notebook`. The file can be a notebook file with a `.nb` extension or a plain text tile with a `.txt` or no extension just as long as the file is written in such a way that Mathematica can understand it.

Appendix F

Mathematical Software Tasks

To become familiar with MATLAB, Maple, and Mathematics, carry out the following tasks. When successful, record your work by saving it and printing it.

F.1. For the function $\tan^2 x$, find the Taylor series expanded about zero, up to and including the term x^{10}.

F.2. Find the first prime number greater than $27,448$.

F.3. Find the indefinite integral $\displaystyle\int \frac{dx}{\cos^4 x}$.

F.4. Find the definite integral $\displaystyle\int_0^1 \log|\log x|\,dx$.

F.5. Obtain the numerical value of $\displaystyle\int_0^1 \sqrt{1+\sin^3 x}\,dx$.

F.6. Find the solution of the differential equation $y' + y = 1/(1+e^x)$.

F.7. Define the function $f(x,y) = 9x^4 - y^4 + 2y^2 - 1$. Find the value of $f(40545, 70226)$ as follows.

a. First, compute it is the straightforward way by direct substitution of $x = 40545$ and $y = 70226$ into the definition of $f(x,y)$. In this work, use first six decimal places of accuracy, then seven, eight, and so on up to twenty-four decimal digits of accuracy. *Suggestion:* Write a program and do floating-point operations instead of integer arithmetic, write numbers with decimals such as 9.0, 40545.0, and so forth.

b. Next, establish by means of the elementary algebra capabilities that $f(x,y) = (3x^2 - y^2 + 1)(3x^2 + y^2 - 1)$. Using this formula, compute $f(x,y)$ at the same values as done previously by again using different precision from six to twenty-four decimals digits of accuracy.

Draw suitable conclusions. (In other words, describe what you have learned from this problem.)

Appendix G

Reviews for Exams

Chapter 1. Introduction and Floating-Point Representation Know two forms of Taylor's Theorem; in particular, the expansions of $f(x)$ about c and $f(x+h)$. Know how they can be used and what their remainder terms are.

Understand the normalized floating-point representation of numbers in decimal or binary form. Understand how numbers have to be approximated in getting them into a computer; in particular, understand single-precision and double precision IEEE standard floating-point representation. Understand the basic rounding errors in entering numbers or operating on them arithmetically. Understand loss-of-significance from subtraction and how to avoid it.

Chapter 2. Linear Systems Understand *naive* Gaussian elimination on a system of equations. Be able to show with simple examples that it is unsatisfactory as a general algorithm. Be able to describe the *scaled partial pivoting* strategy. Be able to count the multiplications and divisions needed in executing it. Be able to compose a pseudocode for solving a tridiagonal system of equations without any *pivoting strategy.* (Memorization is not necessary; just understand the principles, and you will be able to derive the algorithm yourself.)

Chapter 3. Nonlinear Equations Understand the Bisection Method. Understand the rationale of Newton's Method and how the formula is derived. Understand *quadratic convergence* and why it is an advantage. Be able to compute a few steps in the Newton Method using a simple function, and without needing a calculator. You should know the Newton's formula $x \leftarrow x - f(x)/f'(x)$. Be able to explain the source of the Secant Method, and know the formula for it. Know its advantages and disadvantages in comparison with the Bisection Method and Newton's Method. What are their rates of convergence? Be able to compute a step or two in these procedures, given a simple function and a place to start.

Chapter 4. Interpolation and Numerical Differentiation Be able to find an interpolating polynomial for a small table of values, using the algorithm given (without a calculator). Know how divided differences are used on such problems. Be able to write a polynomial in *nested* form. Be able to explain the Lagrange method of getting an interpolating polynomial. Explain why we cannot expect an interpolating polynomial to represent an underlying function well. What lesson is to be drawn from the Runge function as an example? Know and be able to use the formulas for the central difference approximations of $f'(x)$ and $f''(x)$. Be able to derive some useful formulas for $f'(x)$ and $f''(x)$ by means of a few terms of a Taylor series. Be able to assess the error involved in these formulas by retaining some additional terms in the Taylor series. Be able to explain Richardson's extrapolation process.

Chapter 5. Numerical Integration Be able to explain and use the lower sums and upper sums to get crude estimates of a definite integral. Explain why one expects the Trapezoid Rule and Simpson Rule to give results that are more accurate than the upper and lower sums. Be able to derive the formula for the Composite Trapezoid Rule and Composite Simpson Rule, starting with the elementary formula for the area of a single trapezoid. Know the error terms for the Trapezoid Rule and Simpson Rule, and how to use them in a simple case. Know the basic idea of the Romberg algorithm and how it depends on the Trapezoid Rule and Richardson Extrapolation. Know the principle of Gaussian quadrature formulas and how one can derive them.

Chapter 6. Spline Functions Understand the general nature of spline functions of degree 0, 1, 2, and 3. Be able to test a given function to see whether it is or is not a spline function of a certain degree. Be able to write down the formulas for a first degree spline function that has knots $t_0, t_1, \ldots, t_n$ and takes prescribed values $y_0, y_1, \ldots, y_n$ at the knots. (This is interpolation at the knots.) Be able to count the number of available parameters and the number of conditions in this case or in the more complicated case of spline functions of degree 2 and 3. You do not need to be able to derive the formulas for quadratic and cubic spline interpolates, but you should have some idea as to why we have to do some mathematical analysis to get the equations leading to the cubic natural spline. Know why we must solve a tridiagonal system to get the cubic natural spline interpolating function for a table of prescribed values (t_i, y_i). Understand the significance of the theorem concerning the optimizing property of a natural cubic spline interpolant.

Chapter 7. Initial-Value Problems Be able to determine the equations involved in applying the Taylor series method to a single ordinary differential equation as well as to a system of ordinary differential equations. Understand the general ideas behind single-step methods such as the second-order Runge-Kutta method, the fourth-order Runge-Kutta method, and the adaptive Runge-Kutta-Fehlberg method. Understand the basic ideas behind multi-step methods such as the Adams-Bashforth-Moulton method.

Chapter 8. More Linear on Systems Be able to explain how one can easily solve the system of equations $Ax = b$, if an LU-factorization $A = LU$ is available. Be able to explain the slightly more general case when P is a permutation matrix and a factorization $PA = LU$ is available. Be able to explain how one obtains an LU-factorization by the Gaussian elimination algorithm (in the simple case when no pivoting strategy is used.) Understand the basic ideas behind iterative methods such as the Jacobi method, the Gauss-Seidel method, and the SOR methods as well as how to apply them to a small systems of linear equations. Know the basic ideas involved in computing eigenvalues using the power method.

Chapter 9. Method of Least Squares and Fourier Series Be able to obtain the normal equations in a linear least squares approximation to both a polynomial and non-polynomial function. Be able to *solve* an inconsistent system of linear equations, $Ax = b$, in the least-squares sense.

Chapter 10. Monte Carlo Methods & Simulation. Understand the basic ideas involved in a numerical simulation using a random number generator. Have a general knowledge of the Monte Carlo technique for finding areas and volumes as well as evaluating complicated numerical integrals.

Chapter 11. Boundary Value Problems Understand the general concepts and ideas in the shooting method and the discretization method for solving a two-point boundary-value problems.

Chapter 12. Partial Differential Equations Understand the finite difference approach for solving partial differential equations. For parabolic problems, understand the Crank-Nicolson method and its different versions. For hyperbolic problems, understand the marching method approach. For elliptic problems, understand how the finite difference approach leads to a large system of linear equations that can be efficiently solved by iterative methods. Model problems such as the heat equation, wave equation, and Laplace equation are important examples in illustrating numerical methods for solving partial differential equations.

Chapter 13. Minimization of Functions Understand the general ideas involved in the minimization of functions whether in the one variable case or the multi-variable case. Understand key concepts such as the gradient vector, Hessian matrix, and the multi-variable Taylor series.

Chapter 14. Linear Programming Problems Be able to transform a linear programming problem into the First Primal Form, the Dual Form, and the Second Primal Form. For a simple linear programing problem, you should be able to sketch the feasible region and solve it by the graphical method. Have a general understanding of the Simplex Method. In solving an inconsistent linear system of equations by either an ℓ_1-problem or ℓ_∞-problem, be able to write the associated linear programming problem.

Appendix H

Lists of Sample Codes

The tables list files containing sample programs based on the material presented in the book *Numerical Mathematics and Computing, Seventh Edition* Ward Cheney and David Kincaid, Cengage, ©2013. The can be viewed and downloaded from the textbook homepages at

`www.cengage.com/math/cheney`

`www.ma.utexas.edu/CNA/NMC7`

They are intended primarily as a learning and teaching aid for use with this book. Also at the homepage is an up-to-date errata list and teaching aids that we have developed or others have been so kind to send us, which include some overhead transparencies.

We believe that these computer routines are coded in a clear and easy-to-understand style but we have intentionally excluded many comment statements so that students will read the code and study the algorithms—they can add comments as they decipher them. These programs are usable on computer systems with appropriate compilers, from small personal computers to large scientific computing machines. However, they do not contain all of the "bells-and-whistles" of robust state-of-the-art software such as may be found in general-purpose scientific libraries. Nevertheless, they are adequate for many small non-pathological problems.

Our Other Books

Numerical Analysis: Mathematics of Scientific Computing, Third Edition David Kincaid and Ward Cheney, American Mathematical Society, ©2002. See Web pages at

`www.ma.utexas.edu/CNA/NA3`

Linear Algebra: Theorey and Applications, Second Edition Ward Cheney and David Kincaid, Jones & Bartlett Learning, ©2012. See Web pages at

`www.ma.utexas.edu/CNA/LA2`

File Name	Description of MATLAB Code
CHP01/	
sineplot.m	Graph of Taylor series partial sums for sin(x) (invoking `s1.m, s2.m, s3.m`)
sqrt_approx.m	Variable precision arithmetic for approximation
format.m	Numbers in different formats
accuracy.m	Numbers with different accuracy
CHP02/	
gauss_elim1.m	Gaussian elimination to solve linear systems
gauss_elim2.m	Gaussian elimination to solve linear systems
CHP03/	
fun_roots.m	Roots of functions or polynomials (invoking `G.m`)
poly_roots1.m	Roots of a cubic polynomial
newton_sys.m	Newton's method for solving a nonlinear system
gauss_newton.m	Newton's method for solving nonlinear systems (invoking `Fun.m`)
fractal.m	Fractal basins of attraction (CEx. 3.2.27)
poly_roots2.m	Roots of a fifth degree polynomial
CHP04/	
newtn_int_poly.m	Newton interpolation polynomial equidistant pts
inverse_interp.m	Inverse Newton interpolation polynomial example
runge_fun.m	Polynomial interpolation for the Runge function
CHP05/	
num_int1.m	Numerical integration of $\exp(-x * x)$ (invoking `f1.m`)
num_int2.m	Numerical integration of $\sin(x)/x$ (invoking `f2.m`)
num_int3.m	Numerical integration of $\cos(2 * x)/\exp(x)$ (invoking `f3.m`)
cex5_2_8.m	Computer Exercise 5.2.8
cex5_2_9.m	Computer Exercise 5.2.9
CHP06/	
spline_plot.m	Plot of a cubic spline curve
spline_sin_plot.m	Plot of a cubic spline curve for sin(x)
CHP07/	
euler.m	Euler's method for solving an IVP (invoking `f.m`)
rk_ode23.m	Runge-Kutta method for solving an IVP (invoking `ode23file1.m`)
rkf_ode45.m	Runge-Kutta Fehlberg method for solving an IVP (invoking `ode45file1.m`)
rk2_ode23.m	Runge-Kutta method, systems of ODEs (invoking `ode23file2.m`)
rkf2_ode45.m	Runge-Kutta-Fehlberg method for systems of ODEs (invoking `ode45file2.m`)

File Name	Sample MATLAB Codes
CHP08/	
ldl.m	LDL factorization
lu_fact.m	LU factorization
eig.m	Eigenvalue example
null.m	Null space example
timing	Timing eigenvalue computation
sng_val_decomp.m	Singular value decomposition of a matrix
cex8x_4_1d.m	Computer Exercise 8.4.1d
mod_power.m	Modified power method
small_eig.m	Small eige value
inv_power.m	Inverse power mehtod
shift_inv_power.m	Shifted inverse power method
CHP09/	
ls_fit.m	Linear least squares fit for polynomials
np_ls_fit.m	Least squares fit for a non-polynomial function
lstsq1.m	Linear least squares example
p_inv1.m	Minimal solution using Pseudo-inverse of matrices
p_inv2.m	Find pseudo-inverse in case of loss in rank
CHP10/	
rand.m	Examples using random numbers
CHP11/	
bvp.m	Two-point boundary-value problem example (invoking `bvpfcn.m, bvpbc.m`)
CHP12/	
heat.m	Heat Equation
par.m	Parabolic Equation
wave.m	Wave Equation
poisson.m	Poisson Equation
fast.m	Fast Equation
helm.m	Helmholtz.Equation
CHP13/	
fmin1.m	Minimizing multivariate functions
fmin2.m	Find local minimum of a function
CHP14/	
lin_prog1.m	Maximize subject to inequality constraints
lin_prog2.m	Minimize subject to inequality constraints
lin_prog3.m	Minimize subject to equality constraints
lin_prog4.m	Minimize subject to inequality constraints

File Name	Sample Maple Codes
CHP01/	
horner_symbolic	Manipulation of Horner polynomial
sin_plot	Graph of Taylor series partial sums sin(x)
taylor_series	Taylor series expansion of functions
sqrt_approx	Taylor series expansion of square roots
convert	Converting numbers
loss_of_significance	Loss of significance in subtraction
accuracy	Computations with various accuracy
CHP02/	
gauss_elim1	Gaussian elimination
gauss_elim2	Gaussian elimination example
band	Banded linear system example
penta	Penatadiagonal example
pentasym	Pentadiagonal symmetric example
CHP03/	
fcn_roots	Roots of functions
newt	Newton's method example
poly_roots	Roots of a 5th polynomial
CHP04/	
newton_interp	Newton interpolation polynomial
runge_fun	Polynomial interpolation for Runge function
derivative	Symbolic and numerical derivative
CHP05/	
num_integ1	Numerical integration of $\exp(-x^2)$
trapezoid_rule	Trapezoid rule for an integral
comp_trap_rule	Composite Trapezoid rule for an integral
num_integ2	Numerical integration of $\cos(2 * x)/\exp(x)$
CHP06/	
cubic_spline1	Plot of natural cubic spline curve
cubic_spline2	Generate and plot cubic spline curve
bernstein_poly	Graph of few Bernstein polynomials
ctrl_pt_curve	Generating curves using control points
CHP08/	
lu	LU decomposition
lufactor	LU factorization
ce8-4-8	Computer Exercise 8.4.8
char	Eigenvalues via characteristic polnomial
null	Null space, eigenvalues/eigenvectors
eigen	Eigenvalues/eigenvectors (Linear Algebra)
Schur	Schur decomposition (Linear Algebra)
Svd	Singualr value decomposition (Linear Algebra)
sng_val_decomp	Singular value decomposition (linalg)

File Name	Sample Maple Codes
CHP07/	
ode0	Numerical solution to an IVP: example 0
ode1	Numerical solution to an IVP: example 1
euler	Euler's method for solving an ODE
ode2	Numerical solution to an IVP: example 2
ode3	Taylor series method (order 4) solving ODE
ode4	Numerical solution to an IVP
adams_mlt_coef	Adams-Moulton formulas
ode_sys1	Analytic/numerical solution systems of ODE
ode-2nd_order	Second order IVP
ode_sys2	Analytic/numerical solution systems of ODE
CHP09/	
lstsq1	Linear least squares example
lstsq2	Linear least squares example
lstsq3	Nonpolynomial least squares example
minimal_sol	Minimal solution random matrix A/vector b
svd_penrose	Penrose properties for pseudomatrix
CHP10/	
rand_num	Generating random numbers
rand	Generating random integers
CHP11/	
bvp	Solving boundary value problems
CHP12/	
heat	Parabolic PDE: heat equation
wave	Hyperbolic PDE: wave equation
ell	Elliptic PDE
CHP13/	
grad_hessian	Calculate gradient vector and Hessian matrix
CHP14/	
lin_prog1	Maximize subject to inequality constraints
lin_prog2	Minimize subject to inequality constraints
lin_prog3	Minimize subject to equality constraints
lin_prog4	Minimize subject to inequality constraints

File Name	Sample Mathematica Codes
CHP01/	
sineplot	Plots of partial sums for sin(x)
sin	Sine series expansion and plot
series	Taylor series expansions
numbers	Examples of numbers in different bases
precision	Machine precision and accuracy
CHP02/	
lin_sys1	Linear system first example
lin_sys2	Linear system second example
CHP03/	
bisect	Bisection examples
newt	Newton's method examples
newt_sys	Newton's method for a system of equations
secant	Secant Method examples
CHP04/	
newt_interp1	Newton interpolation (5 data points)
newt_interp2	Newton interpolation (4 data points)
newt_interp_sin	Newton interpolation ($\sin(x)$ at 10 equidistant points)
inv_interp	Newton inverse interpolation example
runge_fun	Polynomial interpolation Runge function
derivative	Symbolic and numerical derivative
CHP05/	
error_integal1	Numerical integration: error integral
sine_integal	Numerical integration: sine integral
error_integal2	Numerical integration: $\exp(-x^2)$
bernoulli	Bernoulli polynomial
cos-exp	Numerical integration: $\cos(2 * x)/\exp(x)$
cp5-2-8	Computer Exercise 5.2.8
CHP06/	
cubic_spline1	Natural cubic spline example 1
cubic_spline2	Natural cubic spline example 2
bezier	Bezier polynomials example
CHP08/	
lu_factor	LU factorization
char_poly	Characteristic polynomial example 1
null	Null space example
timing	Timing example
eigen	Eigenvalues/eigenvector example
schur	Schur form example
svd	Single Value Decompositon example

File Name	Sample Mathematica Codes
CHP07/	
explicit	Explicit solution of an ODE
practial	Practial ODE example
ode1	Numerical solution to an ODE 1
ode2	Numerical solution to an ODE 2
ode3	Numerical solution to an ODE 3
ode4	Numerical solution to an ODE 4
ode_sys1	Analytic/numerical solution systems of ODE 1
ode-2nd_order	Second order IVP
ode-high_order	High order IVP
ode_sys2	Analytic/numerical solution systems of ODE 2
CHP09/	
ls_fit	Linear Least squares fit for polynomials
np_ls_fit	Least squares fit a non-polynomial function
cheby	Plot of Chebyshev polynomials
p_inv1	Minimal solution using pseudoinverses
p_inv2	Find pseudo-inverse in case of loss in rank
CHP10/	
ran_num_exps	Examples using random numbers
CHP11/	
bvp1	Two-point boundary value problems example 1
bvp2	Two-point boundary value problems example 2
CHP12/	
heat	Hyperbolic PDE: solve heat eqn. & plot
wave	Parabolic PDE: solve wave eqn. & plot
CHP13/	
min_val1	Minimizing multivariate functions
min_val2	Find local minimum of a fucntion
taylor_series	Taylor series in two variables
CHP14/	
lin_prog1	Maximize subject to inequality constraints
lin_prog2	Minimize subject to inequality constraints
lin_prog3	Minimize subject to equality constraints
lin_prog4	Minimize subject to inequality constraints

File Name	**Sample Fortran90 Codes**
CHP01/	
first.f90	First programming experiment
pi.f90	Simple code to illustrate double precision
oct.f90	Print in octal format
hex.f90	Print in hexadecimal format
numbers.f90	Print internal machine representation of various numbers
xsinx.f90	Example of carefully programming $f(x) = x - \sin x$
CHP02/	
ngauss.f90	Naive Gaussian elimination to solve linear systems
gauss.f90	Gaussian elimination with scaled partial pivoting
tri.f90	Solves tridiagonal systems
penta.f90	Solves pentadiagonal linear systems
CHP03/	
bisection.f90	Bisection method
rec_bisection.f90	Recursive version of bisection method
newton.f90	Sample Newton method
secant.f90	Secant method
CHP04/	
coef.f90	Newton interpolation polynomial at equidistant pts
deriv.f90	Derivative by center diff./Richardson extrapolation
CHP05/	
sums.f90	Upper/lower sums experiment for an integral
trapezoid.f90	Trapezoid rule experiment for an integral
romberg.f90	Romberg arrays for three separate functions
rec_simpson.f90	Adaptive scheme for Simpson's rule
CHP06/	
spline1.f90	Interpolates table using a first-degree spline function
spline3.f90	Natural cubic spline function at equidistant points
bspline2.f90	Interpolates table using a quadratic B-spline function
schoenberg.f90	Interpolates table using Schoenberg's process

File Name	Sample Fortran90 Codes
CHP07/	
euler.f90	Euler's method for solving an ODE
taylor.f90	Taylor series method (order 4) for solving an ODE
rk4.f90	Runge-Kutta method (order 4) for solving an IVP
rk45.f90	Runge-Kutta-Fehlberg method for solving an IVP
rk45ad.f90	Adaptive Runge-Kutta-Fehlberg method
taylorsys1.f90	Taylor series method (order 4) for systems of ODEs
taylorsys2.f90	Taylor series method (order 4) for systems of ODEs
rk4sys.f90	Runge-Kutta method (order 4) for systems of ODEs
amrk.f90	Adams-Moulton method for systems of ODEs
amrkad.f90	Adaptive Adams-Moulton for systems of ODEs
CHP11/	
test_random.f90	Compute, store, and print random numbers
coarse_check.f90	Coarse check on the random-number generator
double_integral.f90	Volume of a complicated 3D region by Monte Carlo
volume_region.f90	Numerical value of integral via Monte Carlo
cone.f90	Ice cream cone example
loaded_die.f90	Loaded die problem simulation
birthday.f90	Birthday problem simulation
needle.f90	Buffon's needle problem simulation
two_die.f90	Two dice problem simulation
shielding.f90	Neutron shielding problem simulation
CHP11/	
bvp1.f90	Boundary value problem by discretization technique
bvp2.f90	Boundary value problem by shooting method
CHP12/	
parabolic1.f90	Parabolic partial differential equation problem
parabolic2.f90	Parabolic PDE problem by Crank-Nicolson method
hyperbolic.f90	Hyperbolic PDE problem by discretization
seidel.f90	Elliptic PDE solved by discretization/Gauss-Seidel

File Name	Sample C Codes
CHP01/	
first.c	First programming experiment
pi.c	Simple code to illustrate double precision
xsinx.c	Example of programming $f(x) = x - \sin x$ carefully
CHP02/	
ngauss.c	Naive Gaussian elimination to solve linear systems
gauss.c	Gaussian elimination with scaled partial pivoting
tri.c	Solves tridiagonal systems
penta.c	Solves pentadiagonal linear systems
CHP03/	
bisection.c	Bisection method
rec_bisection.c	Recursive version of bisection method
newton.c	Sample Newton method
secant.c	Secant method
CHP04/	
coef.c	Newton interpolation polynomial at equidistant pts
deriv.c	Derivative by center diff./Richardson extrapolation
CHP05/	
sums.c	Upper/lower sums experiment for an integral
trapezoid.c	Trapezoid rule experiment for an integral
romberg.c	Romberg arrays for three separate functions
rec_simpson.c	Adaptive scheme for Simpson's rule
CHP06/	
spline1.c	Interpolates table using a first-degree spline function
spline3.c	Natural cubic spline function at equidistant points
bspline2.c	Interpolates table using a quadratic B-spline function
schoenberg.c	Interpolates table using Schoenberg's process

File Name	Sample C Codes
CHP07/	
euler.c	Euler's method for solving an ODE
taylor.c	Taylor series method (order 4) for solving an ODE
rk4.c	Runge-Kutta method (order 4) for solving an IVP
rk45.c	Runge-Kutta-Fehlberg method for solving an IVP
rk45ad.c	Adaptive Runge-Kutta-Fehlberg method
taylorsys.c	Taylor series method (order 4) for systems of ODEs
rk4sys.c	Runge-Kutta method (order 4) for systems of ODEs
amrk.c	Adams-Moulton method for systems of ODEs
amrkad.c	Adaptive Adams-Moulton method for systems of ODEs
CHP10/	
test_random.c	Example to compute, store, and print random numbers
coarse_check.c	Coarse check on the random-number generator
double_integral.c	Volume of a complicated 3D region by Monte Carlo
volume_region.c	Numerical value of integral over a 2D disk by Monte Carlo
cone.c	Ice cream cone example
loaded_die.c	Loaded die problem simulation
birthday.c	Birthday problem simulation
needle.c	Buffon's needle problem simulation
two_die.c	Two dice problem simulation
shielding.c	Neutron shielding problem simulation
CHP11/	
bvp1.c	Boundary value problem solved by discretization technique
bvp2.c	Boundary value problem solved by shooting method
CHP12/	
parabolic1.c	Parabolic partial differential equation problem
parabolic2.c	Parabolic PDE problem solved by Crank-Nicolson method
hyperbolic.c	Hyperbolic PDE problem solved by discretization
seidel.c	Elliptic PDE solved by discretization/Gauss-Seidel method

www.ingramcontent.com/pod-product-compliance
Lightning Source LLC
Jackson TN
JSHW060719030826
106136JS00008B/121